TEACHING CRAFT, DESIGN AND TECHNOLOGY

Technology is now one of the foundation subjects of the National Curriculum and children in primary schools will be assessed in it at ages 7 and 11. In this fully revised edition, Peter Williams gives both specialist and non-specialist teachers the help and practical guidance they need to teach and assess Craft, Design and Technology (CDT) throughout the early and middle years.

Peter H.M. Williams is a former County Education Inspector (District) Staffordshire.

ROUTLEDGE TEACHING 5-13 SERIES
Edited by Colin Richards
formerly of the School of Education, Leicester University

ASSESSMENT IN PRIMARY AND MIDDLE SCHOOLS
Marten Shipman

ORGANISING LEARNING IN THE PRIMARY SCHOOL CLASSROOM
Joan Dean

DEVELOPMENT, EXPERIENCE AND CURRICULUM IN PRIMARY EDUCATION
W. A. L. Blyth

PLACE AND TIME WITH CHILDREN FIVE TO NINE
Joan Blyth

CHILDREN AND ART TEACHING
Keith Gentle

MATHEMATICS 5 to 11
David Lumb

TEACHING CRAFT, DESIGN AND TECHNOLOGY

Five to Thirteen

Second Edition

PETER H. M. WILLIAMS

London and New York

First published 1985 by Croom Helm
Second edition by Routledge 1990
11 New Fetter Lane, London EC4P 4EE

Simultaneously published in the USA and Canada
by Routledge
a division of Routledge, Chapman and Hall, Inc.
29 West 35th Street, New York, NY 10001

© 1990 Peter H. M. Williams

Typesetting by Witwell Ltd, Southport
Printed and bound in Great Britain by
Biddles Ltd, Guildford and King's Lynn

British Library Cataloguing in Publication Data

Williams, Peter H. M. (Peter Harry Mervyn), *1926–*
Teaching craft, design and technology : 5 to 13. 2nd ed.
1. Primary schools. Curriculum subjects: Crafts, design &
technology. Teaching
I. Title
372.35

ISBN 0-415-04716-1

Library of Congress Cataloging-in-Publication Data

Williams, Peter H. M.
Teaching craft, design, and technology, five to thirteen/
Peter H.M. Williams.—2nd ed.
p.   cm.—(Routledge teaching 5-13 series)
Includes bibliographical references (p.).
ISBN 0-415-04716-1
1. Industrial arts—Study and teaching (Elementary)—Great
Britain. I: Title. II. Series.
TT166.G7W55   1990
372.5′044′0942—dc20
                                                    90-34956
                                                    CIP

CONTENTS

List of figures and table   vi
Foreword   ix
Editorial preface   xi
Acknowledgements   xiii

1. What is CDT?   1
2. The role of Craft, Design and Technology   15
3. Problem-solving activity   31
4. Technological awareness   60
5. Communication skills   85
6. The working environment   97
7. Resources for learning   117
8. Assessment and evaluation   124
9. Health and safety considerations   136
10. The way ahead   140

Appendices

A. Tool and equipment schedules for CDT   153
B. Useful addresses   161
C. Suppliers of materials and equipment   163
D. Useful further reading   167

Index   173

FIGURES AND TABLE

FIGURES

1.1 Playground cleaning machine   3
2.1 Aids to learning   29
3.1 The problem-solving process   34
3.2 The design line   35
3.3 An example of problem-solving   38
3.4 Two views of the design process   41
4.1 The ferment of CDT   68
4.2 Possible fields of interest   80
4.3 Fields of interest: air movement   81
4.4 Fields of interest: industrial archaeology   82
4.5 Fields of interest: flight   83
4.6 Fields of interest: crossing a river   84
5.1 Outline drawing   89
5.2 Simple line alphabet   91
6.1 Bench hook   101
6.2 Liaison between specialist subject areas   106
6.3 Middle school workshop layout   110
6.4 Storage of rod and sheet materials   111
6.5 Storage of short ends   112
7.1 Resource indexing   119
8.1 Part of a possible profile sheet for a 6-year-old pupil relating to CDT   129
8.2 Part of a possible profile sheet for a 9-year-old pupil relating to CDT   130
8.3 Part of a possible profile sheet for a 12-year-old pupil relating to CDT   131

TABLE

10.1 Proposed implementation of National Curriculum   147

FOREWORD

A STATEMENT OF PRINCIPLES CONCERNING
EDUCATION IN CRAFT, DESIGN AND TECHNOLOGY
as issued by CODATA – the Confederation of Design and
Technology Associations)

The span of human abilities is very wide, and the development of our civilization and our culture has depended and still depends upon the exercise of many arts and many skills by men and women.

In this country, however, the development of these arts and skills through education is not well balanced. Throughout our education system there is a strong emphasis on literature, history, science and mathematics, but there is too little emphasis on the development of the arts and skills involved in designing, making and doing.

There is a fundamental flaw in our system of social values whereby high esteem is accorded to those skilled at expressing themselves verbally and in marshalling arguments, while much less esteem is accorded to those who express themselves through designing and making, through co-ordinated effort of mind, of heart and of hand.

There is a pervasive misconception that the useful arts of manufacture are derivatives of science, and that as such they are worthy of less dignity and respect than science itself. As a result, the education of engineers has regrettably come to be regarded as a down-market form of science rather than as a culture in its own right.

These attitudes have a particularly corrosive and undermining effect in a nation which is dependent upon the useful arts of manufacture for its means of earning its living amongst the nations of the world.

We who are involved in various aspects of education in craft, design and technology wish to assert the following:

1 That the processes of designing and making are fundamental amongst the many arts and skills that make up the span of human abilities.
2 That, at the primary level, designing and making are every bit as important as reading, writing and arithmetic.
3 That, at the advanced level, the activity of designing and making is every bit as important as literature, history, mathematics and science.
4 That designing and making are valid, indeed indispensable, forms of education.
5 That, too often in education, contrived situations are presented to pupils to ensure that problems:
   (a) can be solved in a limited time;
   (b) can be tackled by each individual;
   (c) have a predetermined solution;
   (d) can be solved using only the information given.

These are highly artificial circumstances bearing little relation to real life. Designing and making activities introduce young people to the notions that there are different solutions to a problem, that some of them are better than others; that they can co-operate in groups to seek solutions; that all the necessary information may not be to hand and that some may have to be discovered and some discarded.
6 That there is a need throughout the education service for full recognition of these fundamental points so that the ability to design and make will be developed along with other basic human arts and skills.
7 That, in order to achieve these aims, we believe that in every school, every year, every child should be actively engaged in the experiences of designing and making.

EDITORIAL PREFACE

Teaching 5–13 is a series of books intended to foster the professional development of teachers in primary and middle schools. The series is being published at a time when there are growing demands on teachers to demonstrate increasing levels of professional understanding and competence. Although the importance of personal qualities and social skills in successful teaching is acknowledged, the series is based on the premise that the enhancement of teacher competence and judgement in curricular and organization matters is the major goal of pre-service and in-service teacher education, and that this enhancement is furthered, not by the provision of recipes to be applied in any context, but by the application of planning, implementation and evaluation of curricula. The series aims to help teachers and trainee teachers to think out for themselves ways of tackling the problems which confront them in their own particular range of circumstances. It does this by providing two kinds of books: those which focus on a particular area of the primary or middle school curriculum, and those which address general issues germane to any area of the curriculum.

This book focuses on craft, design and technology (CDT), a recent candidate for inclusion in the curriculum for the age group five to thirteen. Craft, design and technology is intended to give girls and boys skills in the identification and solution of practical problems, including the design and construction of devices which perform practical functions. The moves towards establishing craft, design and technology as a central concern in the primary and middle years curriculum are in their infancy; the next few years should see clearer expectations established as to the skills and capabilities to be developed, the range of activities to be included in programmes of work and the material provision required to sustain this work. In all these respects this book makes an important contribution to thinking and practice. Its clear practical approach will contribute to the development of what Sir Keith Joseph terms 'versatile competence', both in the children themselves and in their teachers.

Colin Richards

ACKNOWLEDGEMENTS

The author gratefully acknowledges the help and advice so freely given and especial thanks are due to Dennis Pegg and John Fairclough for their constructive reading of the first draft, to Colin Richards (series editor) and to Peter Sowden, Mari Shullaw and Emma Waghorn at Routledge for their guidance and support, to the Association of Advisers in Craft, Design and Technology, the Confederation of Design and Technology Associations and the West Midlands Branch of the British Association for the Advancement of Science for permission to use material originated by them, and to his wife, Mary, for her patience and support.

Special thanks are due to Mr A. T. Gordon for his advice and guidance in the preparation of this revised edition, especially with reference to the introduction of the National Curriculum.

Chapter One

WHAT IS CDT?

Technology is a creative human activity which brings about change through design and the application of knowledge and resources. Technology involves drawing on human knowledge and resources in order to make things work, control things, and alter the way they work. It is a means whereby mankind makes progress and society develops.

(National Curriculum Report: Science for ages 5 to 16: DES 1988)

INTRODUCTION

In September 1980, Professor David Keith Lucas wrote the following words in his foreword to the Design Council Report.

If education is concerned with trying to fit children to play a full part in the adult life of tomorrow, there can be few more important educational experiences for the children than to grapple with the sort of problems they will meet as adults – problems of the environment, of man-made things, and how they can be improved, of the quality of living – or, in other words, 'design' in all its forms. As such, design education is the concern of all boys and girls, not just those who might eventually go on to design-related adult occupations.

Education in design can also be justified on the grounds that good design is crucial to the national economy.

In this country the tradition in education has been to accord much higher priority to the pure sciences and mathematics than to the applied or practical arts. This is neither the way to attract the most intelligent pupils into design, nor does it

1

provide pupils in general with the sort of skills that they will need to practice as adults.

It is not a situation that can be put right inside the schools alone. Society as a whole must decide what it wants for education and fix its priorities accordingly.

Any investigation of the average secondary school syllabus in the early 1970s would have found little evidence of today's subject called Craft, Design and Technology. Schools catering for boys would have timetabled the subject of Technical Studies (perhaps called Handicrafts, Boys' Crafts or even Heavy Crafts) but the approach would have been formal and based on instruction in the acquisition of skill. In the primary school the subject would have been non-existent. However, at long last, thanks largely to the influence of the Schools Council-based project in Design led by Keele University, and the project in School Technology which was inherited by the National Centre for School Technology at Trent Polytechnic from Loughborough College of Higher Education, the scene has changed enormously. With educational content to the forefront, and with no loss in standards of skill or craftsmanship, the subject is recognized as having relevance and importance for all age and ability levels, the content and teaching styles have probably undergone more changes than any other area of the curriculum, and the importance of the subject activity is at last becoming recognized by government, industry and society as a whole. Of course there is still a long way to go. Many primary school teachers still believe that in-service courses on design or on technological awareness or skill are not their concern but, gradually, the scene is changing, and it is certainly hoped that readers of this book will help others to appreciate the significance of CDT in what is possibly becoming a post-industrial society.

Middle schools have certainly played a significant part in helping to convince teachers that relatively young children can carry out quite involved technological activities, and that their lively, enquiring minds can tackle problem-solving activity unhampered by the preconceived ideas which make original thought so difficult for older people (see Figure 1.1).

The provision of properly equipped workshops in the 9 to 13 middle schools, staffed by specialist teachers of CDT, has also

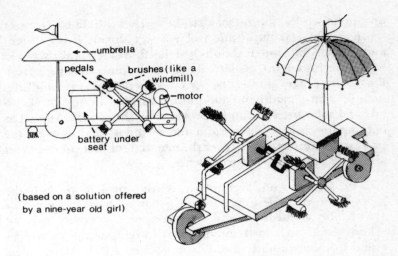

Figure 1.1 Playground cleaning machine

proved that, provided that the problem-solving activity is chosen carefully, very high standards of workmanship can be obtained, coupled with considerable intellectual endeavour. The broadening of the timetable to enable girls to benefit from what were previously considered to be male preserves has also helped to remove teacher prejudice and has opened doors to personal and career prospects which were formerly denied to half the school population.

Although educational change is normally slow and ponderous and quite rightly jealous of its standards and traditions, encouragingly rapid advances are being made in the introduction of electronic and computer-based learning methods. In the average junior school, however, the child with the digitial wrist watch who goes home to play Star Wars on his home computer, and who hopes to get a microprocessor-controlled train set for a Christmas present, still experiences teaching styles and content which would have been familiar to his grandparents. Many schools are failing to adapt to a rapidly changing environment and yet, if children are to be equipped to take their places in society, it is essential that all of them should be helped to acquire technological awareness from the earliest days of schooling. This does not mean the introduction of a new subject in the primary or middle school, but rather that the teacher should be aware of the technological implication of what is already being taught and that lesson content should be related, where appropriate, to today's technological society. At the same time all pupils from the earliest age should be given experiences in handling, manipulating, modifying and joining as wide a range of materials and substances as can be provided by the school. As well as helping the child to observe and understand its environment, the process of gaining control over the materials of that environment also gives enormous self-confidence.

Craft, Design and Technology is concerned more with the development of desirable attitudes than with an end result or with the acquisition and retention of a specific body of knowledge. Its added commitment to design-based activities and to the exploration and exploitation of technology offers further opportunities for the development of altruistic attitudes. The work should also promote an active and informed attitude towards environmental development.

(Curriculum 11–16, Her Majesty's Inspectors, 1977)

This emphasis on attitudes and awareness in the field of design and technology means that the process is closely connected with cross-curricular activity and that all teachers, whatever their subject interests and age-range responsibilities may be, have a duty to exploit the technological implications of lesson content in order to help their pupils to acquire technological awareness. For most teachers this will not be easy because the traditional educational pattern in Great Britain has meant that those teachers have themselves been denied access to technological and design-based influences. Very few initial teacher training courses in Institutes or Faculties of Education contain any reference as yet to this important aspect of general education resulting in almost total technological illiteracy amongst the teaching profession, and therefore self-help and locally planned in-service courses must be relied upon to provide the expertise to enable the teaching force to cope with the needs of twentieth century society.

There is confusion, too, because of the lack of understanding of the real meaning of the terms used. The title 'Craft, Design and Technology' uses words which have their own meanings when used separately: meanings which vary according to the context in which they are used and, although the title does describe the activity clearly to the specialist CDT teacher, it would appear to give little guidance to the uninformed. The average reader is unsure about the differences between tradesmen, craftsmen, mechanics, technicians, engineers and technologists; between artists and designers; even between science and technology. In West Germany the word 'technik' is used to describe the subject area under discussion, but in Great Britain no similar word exists and many chapters have been written in Design Council and other publications to attempt an accurate and unambiguous description of the range of activities encompassed by the title CDT. Perhaps the word ingenuity best describes certain aspects of the process, but even this conjures up mental pictures of Heath Robinson-type contraptions or Emmett masterpieces neither of which fit the bill. The mass media, too, frequently refers to scientists when the credit should have been given to technologists, as the scientist's role is in research and understanding whereas the role of the technologist is in the application of knowledge and resources to satisfy a human need.

It is hoped that the description of the processes which together make up the activity known as Craft, Design and Technology which follow in this book will help the reader to join the ranks of the well informed and to assist in the task of developing the curriculum to cater for the possible needs of pupils who will live most of their lives during the twenty-first century.

The Education Reform Bill enacted in 1988, and introducing a National Curriculum, will result in all pupils studying Design and Technology throughout their period of compulsory education. In the early 1990s many teachers in the primary phase will find some difficulty in achieving a reasonable level of self-confidence in this, to them, new subject area, but this book, together with the planned INSET courses, should help them to attain appropriate levels of competence.

DEFINITIONS

The interrelationships between Craft, Design and Technology

It may help to consider each of the aspects of craft, design and technology separately in order to arrive at a composite definition which truly describes this area of the curriculum. Unfortunately each of the words in the title can be used for a variety of meanings depending on the context in which they appear and the attitudes and philosophy of the reader or listener. A typical dictionary definition would probably contain some, or all, of the following:

craft: dexterity; skill; skilled trade; an art;

design: a preliminary sketch; plan in outline; a plan or scheme formed in the mind; project; intention; sketch plan; relation of parts to the whole; arrangement of details; disposition of forms and colours; decorative pattern; geometric pattern;

technology: science and history of the mechanical arts; a systematic knowledge of the industrial arts; a discourse or treaty on the arts; scientific nomenclature; an explanation of terms used in the arts.

A craftsman could be described as 'one engaged in the crafts', a designer as 'one who produces designs or patterns, a plotter', and a technologist as 'one skilled in technology'. The average person, however, would need definitions of terms such as skilled or unskilled before he could evaluate a craftsman's status. He would almost certainly think of patterns or the application of a transfer-based decoration when faced with the word 'design', and to him a technologist would be synonymous with an engineer or a mechanic, both wearing greasy overalls and carrying an oilcan and a bunch of cotton waste in either hand. No wonder the relatively new title of craft, design and technology means so little to the majority of primary school teachers.

Almost certainly the term CRAFT is the easiest of the three to define as the issue is not confused by ambiguity, and just about everyone understands what is meant by craftsmanship even if there is little understanding of the relationship between hand- and machine-tool skills and methods. CRAFTMANSHIP is concerned with the development of manipulative skills based on making and doing. This means that it is about the acquisition of skill and the application of techniques in order to produce well-designed artefacts. These skills and techniques can be acquired only if there is also an understanding and familiarity with the materials being used and the end result is a mastery or control over them.

Unfortunately many teachers still believe that skills must be acquired as a separate, initial stage, divorced from and preceding the design/make/evaluate process. In the world of industry this separate method is probably the best as apprentices and others are motivated by the desire to reach certain standards and their skills base is wide enough to enable them to see the purpose of the various training exercises. Until recently this was akin to the process applied by the majority of handicraft teachers in schools. Originally, a very limited version of the subject was taught by drill methods with numbered stages and operations and set procedures, but this gradually evolved into a more flexible system, still skill based, but where the exercises were disguised as 'jobs', each pupil aiming at an identical product, skills and techniques being demonstrated stage by stage, and the whole process being rather like painting by numbers. Sometimes the pupils worked from orthographic drawings, but more usually each stage was introduced by the teacher from a blackboard drawing or even with no drawing at all.

The advantages of this system were that organization was easy, materials could be prepared or cut to size in advance, skills could be demonstrated to the whole class and, in theory, the rate of progress could be controlled. In practice, of course, there were many disadvantages. The pupils worked at different rates resulting in some pupils getting further behind at each demonstration, with a longer and longer time-lag between seeing the process demonstrated and then carrying out the technique on their own jobs, accompanied by an inevitable drop in standards. At the other end of the ability range, some pupils would finish the process quickly, and therefore have to waste time waiting for others to catch up to enable enough pupils to be ready for the next demonstration to take place. The teacher had little choice but to pitch the standard around the middle ability range and the better pupils were rarely stretched. The biggest disadvantage of all, however, was that many of the pupils carried out the processes without any real understanding and with little personal involvement or contribution, rather as if a teacher of English did nothing but dictation at every lesson.

The acquisition of skill for its own sake is of little value unless that skill can be applied to some purpose. The golfer may develop a perfect swing but this is of no use unless he has the opportunity to drive a ball up the fairway from time to time.

With craft, too, the need for a particular skill should be

7

established before tuition is commenced, as the resultant motivation will almost certainly guarantee higher standards and will enhance the learning situation.

Because of this change from a teaching to a learning environment, it has become possible to introduce design/make/evaluate processes (problem-solving activity), and the pupils then create needs for specific skills and techniques, without which they could not carry through their own solutions. A skilled teacher will introduce constraints into the design situation so that only a narrow range of new skill is needed, enabling class and group demonstrations to be given, and for adequate follow-up supervision to be practicable.

The discussion now leads us to the second word in the subject title, namely 'design', and as this book is concerned with teaching strategies rather than philosophical concepts the temptation to indulge in long abstract arguments and definitions must be avoided. As the reader will have gathered, it is extremely difficult to give a concise definition of design because of the enormously wide field of activity where the term is used. However, for the purpose of this chapter, a suitable definition might be that design is the development of an appropriate awareness of the aesthetic and functional 'rightness' of man-made objects based on problem-solving activity.

The ability to discriminate between what is good and what is less good is part of the design process and is a vital skill in our consumer-based society, but without a knowledge of tool usage, materials and processes there can be no effective assessment of quality and fitness for purpose. Properly-guided problem-solving activity takes the pupil along the design spiral, leading him from one logical step to the next and demanding a discriminatory approach all the way. If situations and problems are carefully chosen to give both continuity and progression then the acquisition of design skills will be cumulative. Self-evaluation is an important part of this process and is built into the problem-solving approach.

Design skills are needed in both craft and technological activities and they constitute the unifying factor in CDT work.

To simplify matters, there are four main areas inherent in good design:

1 Fitness for purpose.

2 Correct use of tools and materials.
3 Sound construction.
4 Pleasant appearance and suitability for the surroundings.

The fourth factor usually emerges as the direct result of the correct application of the other three.

From this explanation it will be obvious that design activity is at the core of all work in CDT and the layman's concept that design is something which is added afterwards must be eradicated.

The design process is particularly important in technology, the third element in the CDT framework, but, first, what is meant by the term 'technology'?

Once again there is the risk of over-simplification, but a concise definition could be that technology is the development of the ability to make purposeful use of knowledge, materials and resources to satisfy human needs and, in the school context, for any age and ability level, this would be based upon an extended view of the pupil's own environment and understanding. It is therefore that process which is concerned with the individual's reaction and interaction with the environment, and particularly with the man-made elements of that environment which concerns the teacher of craft, design and technology.

In school there are two main aspects of technology which must receive consideration. The first and most fundamental is that all teachers should ensure that there is sufficient technological bias in their subject work and also sufficient cross-curricular activity to guarantee that there is *technological awareness* on the part of all pupils. This area must be the responsibility of all teachers and it is too important to be left to the limited time available to specialist teachers. The second area is concerned with *technological competence* and ability and is more specialist in nature in that CDT activity is directly connected with work in art and in science. Both areas, however, are concerned with mans' ability to satisfy human need as already defined in this chapter.

Perhaps we are now ready to attempt a composite definition to describe this broadly-based subject area called craft, design and technology.

Craft, Design and Technology is concerned with making and doing. It encompasses aesthetic understanding and discrimination and incorporates the acquisition of skills enabling tools, materials, processes and knowledge to be used in the control of the environment and to satisfy human need.

A CURRICULUM STATEMENT

Craft, Design and Technology, as a major curriculum area, has the central aim of developing the intellectual capacity and practical skills of the pupil through the process of designing, making and evaluating.

A principal objective is to encourage pupils to develop skills using resistant materials to solve realistic practical problems. These problems cover a wide range of activities often embracing the resources of science as well as the consideration of aesthetic values. This process requires a progressive accumulation of knowledge, intellectual and practical skill, experience and the ability to communicate in both verbal and graphic terms. The activity of designing and making requires an understanding of human needs and values, and provides opportunities for the realistic application of other disciplines. As such, Craft, Design and Technology embodies the essentials of a truly educational process and should be part of the common experience for pupils of all ages and abilities.

Craft, Design and Technology should provide the opportunity for improving the balance between theoretical and practical aspects of the curriculum. The workshops are a fertile environment where basic theoretical knowledge can be applied and developed and the skills learned there, and elsewhere, reinforced in an atmosphere more closely related to the world outside school. This environment creates an atmosphere where highly-motivated learning is both possible and likely and which fosters the values of both divergent and convergent thinking. These considerations have important implications in the preparation of pupils for Further and Higher Education and for industrial life.

In the primary phase and in the earlier years of secondary education, Craft, Design and Technology should permeate the whole curriculum with an emphasis being placed on the acquisition of technological awareness. This should be consolidated by workshop activities where technological abilities should be acquired and where close relationships must be developed with both scientific and art and craft activities.

In addition to the development of general skills and the acquisition of conceptual knowledge, Craft, Design and Technology is also concerned with the development of social qualities such as patience, tolerance, sensitivity and group co-operation. The value of planning has long been recognized, by teaching pupils to define, to discriminate, to execute to a high standard and, finally, to evaluate the chosen solution. Craft, Design and Technology makes important contributions to individual development. The subject embodies aspects of environmental and consumer education and encourages an understanding of, and sometimes involvement with, industry and commerce. Whilst all these considerations are vital for personal development they are also essential in preparing pupils to take an informed, active and positive role as adults in a rapidly changing technological society.

In the sphere of technology, pupils learn how to use knowledge, materials and resources in order to satisfy human needs and they also have opportunities to make and control devices by mechanical, electronic, microprocessor and computer-based means.

(Based on a statement issued by the Association of Advisers in Craft, Design and Technology in 1982)

AIMS AND OBJECTIVES
Aims

As with all subject areas, it is essential that all teachers of CDT have clearly defined aims and objectives for their work. Of course, these will vary from teacher to teacher and from school to school, but it is hoped that this chapter will assist in identifying those aims and objectives which are important in particular circumstances. Much will depend on the age and ability levels of the pupils being taught, but in addition the school's environment, the facilities available and the experience and expertise of the teachers involved will all affect the situation, particularly where objectives are concerned.

Perhaps an attempt at defining the differences between aims

and objectives would not come amiss at this stage because, although the two are interdependent, it is generally the aims which appear in the scheme of work and the objectives which set the scene for lesson preparation notes.

Aims generally refer to the long-term or short-term intentions of the teacher, relating to the general development of the pupil. In general educational terms these would probably have a large area of common ground for all subject activity and could include the following aims.

- To enable the pupil to experience areas of knowledge.
- To develop skilful habits of work and endeavour.
- To enable the pupil to acquire and use knowledge and skill.
- To develop high standards.

In the primary and middle school and in the earlier years of secondary education the general aims set out in the Green Paper of 1977 contain four which are particularly relevant in CDT work.

I To help children develop lively enquiring minds, giving them the ability to question and to argue rationally and to apply themselves to tasks.

IV To help children to use language effectively and imaginatively in reading, writing and speaking.

V To help children to appreciate how the nation earns and maintains its standard of living and properly esteem the essential role of industry and commerce in this process.

VI To provide a basis of mathematical, scientific and technical knowledge, enabling boys and girls to learn the essential skills needed in a fast-changing world of work.

Of course there are aims which are specific to various areas of CDT teaching and the inclusion of any of these in a scheme of work will depend on the bias applied to the teaching of the subject. The following aims are some that would be found in many schools.

- To help the children to acquire understanding and expertise through the processes of designing and making.
- To help children to become aware of man's technological development.
- To provide an opportunity for children to handle and use a wide range of natural and man-made materials.
- To provide opportunities for children to widen their range of

skills through the use of hand and machine tools under controlled conditions.
- To encourage the development of personal qualities of ingenuity, resourcefulness, initiative and self-reliance.
- To provide opportunities for children to acquire skills in mechanical, electrical and electronic control.

Obviously before committing ourselves to subject activity we must be absolutely clear in our own minds why it is that we intend to involve the children in that activity. Hence clearly defined aims are essential for every teacher and these should be to the forefront of the teacher's thinking.

In giving a list of aims appropriate to the age range covered by this book it must be understood that the list is not exhaustive, neither do the aims listed apply equally to all schools and to all teachers. It is offered purely for guidance purposes.

Some aims for CDT teaching for the 5–13 age range

- To provide an opportunity for children to handle and use a wide range of natural and man-made materials.
- To widen the child's knowledge of the immediate environment and to give increasing control over that environment.
- To allow the individuality of the child to develop through the inventive and creative possibilities which the work should offer at every stage.
- To provide a basis for concrete thought and eventually to create a base of success from which linguistic skills may develop.
- To help the progressive development of motor skills.
- To help towards the development of logical thought by using design processes as this approach has relevance throughout life.
- To give a sense of satisfaction and achievement without emphasizing the competitive element.
- To help to develop standards and the beginnings of aesthetic awareness together with the ability to exercise discrimination.
- To help provide the common ground necessary for the correlation of subject activities.
- To encourage leisure-time interests.
- To encourage the willingness to question and experiment.

- To exploit the senses of sight and touch.
- To develop attitudes of self-discipline, independence and an acceptance of responsibility.
- To develop tolerance, consideration for others and the ability to work as a team.
- To develop the ability to plan ahead.
- To develop a positive concern for safety based on common sense.
- To encourage inventiveness, resourcefulness, ingenuity and imagination.
- To develop the ability to retrieve information.
- To develop high personal standards.

Objectives

Objectives are concerned with content and with what the individual child will actually achieve by doing the work. The following are examples of objectives.

- To make simple working models of artefacts found in Roman life.
- To show examples of basic industrial processes.
- To be able to evaluate a chosen solution to a problem.
- To produce a working drawing of the chosen solution.
- To revise the method of cutting and filing Perspex.
- To paint and display the finished models.

With young children short-term objectives are very necessary to help sustain interest and endeavour, and lesson preparation should always try to identify short-term targets of this type. This approach is particularly effective with less able children.

Chapter Two

THE ROLE OF CRAFT, DESIGN AND TECHNOLOGY

Much of children's early technological experience comes from solving problems and responding to the needs created by their imaginative play. . . . These activities provide a range of experiences which can be harnessed and developed by the teacher within the classroom.

(National Curriculum Report: Science for ages 5 to 16: DES 1988)

THE EARLY YEARS PHASE OF EDUCATION

Prior to 1982 it was unusual to hear any reference to CDT as an activity in the primary or first school curriculum, but during that year the BBC introduced its very first Junior Craft, Design and Technology programme, 'Up and Down the Hill' and teachers began to realise that perhaps this subject area did have relevance in the early years phase after all. For many years the local educational advisers and inspectors as well as Her Majesty's Inspectors (HMI) had been promoting the need for teachers of primary school children to become involved in problem-solving and technological activity, but success was very patchy, relying on a few local enthusiasts to develop activity in their own schools whilst most other teachers maintained that it was not for them as they were 'no good with their hands', or that 'they couldn't draw for toffee', or that 'it is not a primary school subject'. The real problem appears to have been that early years teachers were put off by the subject title and this is where the television screen, with its ability to use graphics and pictorial situations, was able to

15

score some success by helping to convince teachers that the activity portrayed being carried out by junior and middle school children was appropriate to their own situations.

In the programmes pupils were encouraged to use the more resistant materials of wood, metal and plastics in a fairly simple way in order to create practical solutions to the problems raised in the series; to make verbal and visual observations and records and to tackle problem-solving activities in real-life situations. With work of this type it is essential that there should be positive links with activities in language development, mathematics, science and environmental studies and with other areas of the curriculum where appropriate.

Other initiatives will doubtless follow, but at the time of writing it would appear that one hurdle has been cleared and that early years teachers are beginning to recognise that Craft, Design and Technology is an activity which should be part of the curriculum for pupils of all ages and ability levels.

Because we are concerned not with a new subject but simply with doing things more comprehensively and therefore better, it should be understood that there is no need for elaborate equipment or additional facilities to enable CDT activity to take place in the first or primary school. As with any art and craft work, it is desirable to provide easily cleaned and firm working surfaces for practical craft work but this is purely an extension of existing craft activity. In this area pupils should be encouraged and enabled to handle and explore as wide a range of materials as is possible within the facilities and resources of the school. However, the criticism contained in the DES report 'Primary Education in England' (HMSO 1978) that 'more discrimination in the selection of materials by teachers . . . would have contributed to better standards of work . . .' and that 'The emphasis which has been placed on children using a wide variety of materials has in some cases resulted in children working in a superficial way . . .' should be borne in mind. What is needed is guidance to enable the pupil to select those materials most relevant to the problem in hand, and to concentrate on that very limited range with other materials introduced over the months and years in order to achieve control and understanding linked with progression and high standards of skill. At times the most suitable materials will be papier mâché or cardboard . . . at others it will be necessary to introduce galvanised wire, drinking straws and strips of wood if good solutions are to be found and evaluated. To let the children loose with a confusing selection of 'junk' materials may be appropriate for certain types of art and craft activity but it will not lead to the logical approach to problem-solving processes which CDT activity demands. Of course, as with the adult world, at times a child will come up with an intuitive or spontaneous solution to a problem and this must be accommodated if the idea appears to be sound. It is very important, however, to ensure that the activity leads to a successful conclusion no matter whether the approach was logical or intuitive.

The first two years of schooling

At this stage, of course, manipulative skills are very limited, but infant children enjoy handling materials and they have constructional skills which can be exploited. Sand and water play which are recognized as having mathematical connections also have technological relevance. The behaviour of the two materials can be observed, the fact that dry sand can be 'poured' just like water is important, and the discovery that damp sand can be used for moulding sand pies and castles should not pass unnoticed. Building with bricks is a technological activity and the child can be helped to see why some structures will remain standing whilst others topple over. Most infant children learn how to arrange bricks to bridge gaps fairly quickly but it took early man many thousands of years before he learned how to place a log or slab of stone across others to bridge a stream or river. Very young children have neither the strength nor the manual dexterity to manipulate rigid materials but they can learn spacial relationships using wooden and plastic educational toys and they can knock wooden or plastic dowels through holes in exactly the same way that a hammer and nails would be used. Large scale plastic constructional toys provide the opportunity to learn how to connect things together to create new shapes and ideas. The teacher's main role in all this, apart from creating the environment which enables constructive play to take place, is to help the children to observe relationships, causes and effects, textures and surface patterns; to introduce new words and phrases relating to the objects, materials or physical sensations encountered and, above all, to structure the work so that there is real progression in the learning sequence.

At this stage, too, it is important to place real value on the children's use of symbols to communicate ideas. In the primary phase generally, the use of drawings to communicate ideas is very much undervalued yet symbols can be used much more precisely than verbal language for certain areas of communication and even more so at a time when the child's vocabulary is extremely limited. The imposition of adult standards of drawing should be avoided at all costs and the child should be encouraged to express ideas and thoughts by means of simple drawings using thick pencils or 'jumbo' wax crayons.

The teacher of infant age children needs a resource of materials relating to the child's immediate environment . . . string, cord, rope, paper, card, wood, metal, powder, liquid, wax, and so on, so that as new words relating to the environment or to physical sensations (smooth, hard, shiny, cold, and so on) are introduced they can be reinforced by physical contact. Similarly, during constructional play every opportunity should be taken to reinforce words acquisition.

Colourful picture books which extend the child's environment should be available so that concepts can be developed. The ability to recognize the many forms which houses, bridges, ships, aeroplanes or cars can take is a technological concept and one which should be extended.

In handling materials they should be helped to understand that some materials will bend whilst others will not. That some can be cut with scissors and others can be torn. That some 'disappear' when placed in water whilst others appear to remain unchanged. That the properties of certain materials can be changed (for example, powder colour by adding water) whilst others remain stable. Naturally, there is no hard and fast dividing line between scientific and technological involvement at this stage and, or course, words used to describe the findings or observations will be appropriate to the age range but, none the less, genuine scientific and technological activity can take place.

Adhesives such as PVA or Marvin, and the modern safe contact adhesives, enable constructional craftwork using wood, card and fabric to take place with small groups or individuals and the educational possibilities arising from this work should be exploited with emphasis being placed on vocabulary, sensual perception, relationships and the idea of structural development.

A later section in this book develops the idea of 'Fields of Interest' and this type of activity has relevance for the infant stage.

Where hand tools are used these must be selected for their suitability for the age range (advice on suitable tools can be found in the section on classroom organization) and the teacher must ensure that the children are taught how to use the tools correctly. A pencil, wax crayon or paint brush is a tool and it is accepted that the children must be taught how to hold and manipulate these items, and yet some teachers, because of their own lack of tool skill, will quite happily leave a child to find out for itself how to use a hammer or a pair of pincers! If no suitable in-service course in basic tool skills can be arranged it should not be too difficult to arrange for some tuition from the CDT staff of your friendly neighbourhood comprehensive or middle school. This in fact could lead to liaison and co-operation between the CDT department and your own school to the advantage of both.

Liaison between the infant stage and the next stage of the educational process is also very important. It is normally taken for granted that there will be transfer of information and records about a child's reading and mathematical ability but how often does discussion about a child's craft or design ability take place? As was stated in a Design Council publication in 1980:

> In our society, verbal and numerical expression have come to be regarded as more important than other vehicles for thought, a point of view that is fostered because such expression lends itself to systematic teaching and relatively straightforward examining. But a great deal of creative thinking in art, science and technology does not involve words or symbols. . . .

Where a child appears to have exceptional skill or ability this should be encouraged and developed. Unfortunately we are better at helping the slow learner than we are at providing challenges for the gifted child and, worse still, if those gifts are in areas other than those covered by verbal and numerical skills then normally they will go unnoticed.

The infant stage should be one of exploration and discovery, of trial and success, and a time of excitement and achievement and, of course, three-dimensional and practical activity has a major part to play in the achievement of these attitudes in the properly organized infant school scheme.

From 7 to 9 years

Increased vocabulary and a greater awareness of the environment make this an exciting stage of development and yet the child meets many frustrations because its ideas and aspirations outstrip its physical strength and abilities:

> It was common practice for teachers to initiate work in art and craft by directing children's attention to arrangements and displays both inside and outside the classroom, and, to a lesser extent, to the immediate environment of the school. Displays of natural objects such as plants, rocks or shells were introduced as a stimulus for work in over three-quarters of the classes although there was seldom sufficiently careful observation and discussion before the work was begun. Drawing or modelling from direct observation was rarely encouraged. . . . Man-made objects were less frequently used than natural objects as starting points for work in art and crafts, and the observation of mechanical artifacts was rare.

(Primary Education in England 1978)

For many primary school teachers, the world of science stops at the nature table and the world of technology does not appear to exist! In point of fact, however, the average primary school curriculum contains numerous areas of experience which present the possibility for technological activity, and a slight switch of emphasis, or an extension to the work being undertaken, would satisfy the need without any suggestion that a new subject was being added to the curriculum. The most fruitful area at this stage will be found in the work undertaken under the subject headings of Environmental Studies or History and Geography, where the subject content is often concerned with man's interaction with his environment. Man's historical development has been very closely related to his technological achievements, which have been sequential in nature, and these achievements have affected his ability to exploit his environment. If the teacher can try to recognize the technological relationship inherent in the work already being undertaken by the children, then it is not very difficult to increase the emphasis on this aspect of the work with resulting benefits.

Perhaps the process will become clearer if one or two practical examples are discussed with particular reference to the 7 to 9 years age range, although it must be understood that it is necessary for the teacher to be selective and to try to exploit the interests of the pupils. From the lists which follow it may be possible to investigate only a few of the potential topics in the classroom, and the decision whether to tackle one or two of them in depth or to look at a number in a more superficial way must be made objectively by the teacher to suit the immediate needs of the pupil or group. There appears to be little doubt that the best plan of action under normal circumstances would be to select a very limited number of aspects to be investigated by small groups with a larger number of interesting points being discussed at class level in order to set the scene and to broaden understanding.

A very common activity for this age range under the heading of 'geography' is for the children to map their immediate environment, starting with the layout of the school and then working outwards to include the surrounding roads and streets and their own homes. This activity of mapping is itself technological in nature as symbols are being used to denote real objects and places and the ability to create a simple map represents an enormous conceptual advance. The process of visual communication is discussed in greater depth in a later chapter but it must be said here that far too little attention is afforded to mapping, drawing and sketching in the majority of primary schools, and it is certainly very necessary to teach skills of layout, use of symbols and an elementary introduction to scale if the resulting map work is to be of educational value.

In the process of mapping the school and its environment the possibility of discussing roads, bridges, transport, lighting, drainage, pollution, houses, concrete, tarmacadam, gradients, distances, telephone wires, front gardens, walls or fences may arise under skilful leadership and questioning by the teacher, and it should be obvious that each of these topics comes under the technological umbrella. During the discussions, pertinent words should be listed and emphasized so that they are added to the children's vocabulary.

Perhaps transport is one of the topics selected from the possibilities for further development. Reference books and magazines with plenty of illustrative material would be needed to enable the children to research the topic and it could perhaps be combined with a traffic count to establish current transport patterns. Present day modes of transport could be compared and contrasted with those in use, say, 100 and 500 years ago. The materials used in a modern car, motor cycle or pedal cycle could be listed against those used in a vehicle of the periods chosen and the reasons for the changes in materials could be discussed. Most schoolteachers would expect their children to draw or paint pictures of the topic but it is also important for model-making to take place, perhaps using cardboard boxes, polystyrene ceiling tiles or offcuts, modelling clay and wire. The problems of adding wheels to the models would be an important part of the activity and, with some pupils, could even lead to questions about bearings and friction although, for most, the problem of adding wheels which will revolve will be a big enough problem. Some children will want to make the vehicles 'go' and here, power sources based on cotton reels, rubber bands and dowel rod or pencils may be adequate. The model vehicles could be based on Lego or similar construction kits, perhaps with an added body made from cardboard cartons and, in this case, the wheels would present less of a problem although it would be important to encourage the children to discover how the wheels and axles work. If this work could be enhanced by a visit to an appropriate museum, or perhaps by the use of film or video, the historical connections would be that much more vital. It is important, too, to consider the people who use the vehicles. What sort of clothes were worn 100 and 500 years ago? Could people wearing those clothes get into, say, a present day Mini car? What sort of people use the various forms of modern day transport? Who used transport and for what reasons in the earlier periods chosen?

If more than one 'powered' vehicle is made, try running them against one another to see which goes furthest. Why does one travel further than the other? What could be done to improve the performance of both vehicles? Will either of them go uphill? Do they travel faster if they are going downhill? These and similar questions should be raised and possible solutions found. Indeed it may even be possible to consider the problems as to why a bus does not topple over sideways, by using an empty cornflakes packet and one partly filled with sand for comparison, and thus to introduce the principle of centres of gravity.

Another area chosen for investigation from the list suggested may be 'bridges'. The group could be encouraged to trace the history of bridge building techniques from the earliest beam types to the modern box girder constructions making drawings and simple models of some of them. In doing this questions should be encouraged about the various materials used, where they came from, why bridges are needed, the differences between pedestrian, vehicular, railway and canal bridges, why there are very few toll bridges still in operation and so on. If it is not possible to take the children to look at examples of real bridges then many of the constructions used in bridge building can be found in the school. Look at archways, door and window openings, the school hall roof and similar details for a comparison with bridge constructions. The pupils should also be expected to experiment with various materials and sections in an attempt to bridge the gap between, say, two bricks. By placing various materials between the bricks it should be observed that some are more suitable than others, perhaps by rolling a small model car across the 'bridge'. This could lead to an experiment where a plastic seaside bucket is suspended from the centre of each 'bridge' and gradually filled with pebbles until the material starts to give way. The results could be tabulated and then steps could be taken to 'improve' the materials being used, perhaps by folding or rolling them or increasing the thickness and, with some pupils, it may be possible to discover the principles of triangulation and to start to apply this knowledge to a solution of the problem.

Children of primary school age are naturally curious about their environment and it is part of the teacher's responsibility to encourage that curiosity whilst ensuring that the resources are to hand to enable the children to discover possible answers to their questions at a level appropriate to their age and ability. Some teachers believe that they must know all the answers themselves, hence their reluctance to move away from a purely didactive teaching method where the pupils do no more than reflect or regurgitate the information presented to them, but most teachers nowadays realize that a mixture of didactic and exploratory methods gives the best results. Some things have to be taught, particularly if they are skill or technique based or if there is insufficient time or educational purpose for discovery methods, but others are best assimilated by child-centred activity. However, no matter what method is used, the teacher must have targets in view so that each child is being taken forward towards new areas of knowledge and experience.

In all of this the children's written work should be based on their own discussions, observations, discoveries and constructions. It is a complete waste of time and energy for them to copy chunks out of their reference books when they could be recording their own experiences.

Most teachers would hope to be successful in helping children to improve their powers of observation and understanding and, of course, work in art and craft plays an important role in this, but few teachers of the 7 to 9 years age range emphasize the sense which is most highly developed in this age range and which in fact starts to deteriorate shortly after this age, namely the sense of touch. The children should be given opportunities to handle a variety of materials and objects, recording their experiences whenever possible. Fabrics such as velvet and fur fabric, objects made from wood, metal and plastics should be examined and sensations compared with similar experiences with natural things such as stones, plants and fruit. Careful observation of man-made objects such as gearwheels, the handle on a pencil-sharpener, the cap of a ball-point pen, the buckle on a bag or shoe, the shape of a pair of scissors and other familiar objects should be consolidated by simple drawings which should be seen as an alternative to written language.

Craftwork using wood, metal and plastics should play a bigger part in schemes for this age range as there is a marked improvement in the ability to handle tools at this stage. The wood used should be fairly soft and knot free and both Jelutong and Lime are particularly suitable. Balsa wood could be used but it is expensive and its porous nature presents limitations in its use although the very wide range of sections available does make it useful for bridge building and similar constructions. Models made in connection with other subject activity should incorporate wood and thick card, wire can be used as a modelling material and filing of acrylic resin (Perspex) shapes can be attempted provided that a vice or clamping device is available.

THE MIDDLE YEARS PHASE
From 9 to 11 years

Improved physical dexterity and strength and the firmly held belief that nothing is impossible characterizes this age grouping and presents the teacher with a considerable challenge if enthusiasm is to be maintained and standards raised throughout the two years. Children of this age range are to be found in primary, middle (8–12) and middle (9–13) schools, all with different facilities and teaching strengths, and yet the needs of the pupils are the same no matter what type of institution they may be in. It would indeed be most beneficial if teachers of this age group working in areas where there is a variety of educational provision could spend some time visiting colleagues in other types of school in order to be able to compare the attainments of similar age groups in different situations.

Children of this age range attending 9–13 middle schools would appear to be at an advantage in the area of craft, design and technology because, unlike their opposite numbers in other types of school, they have access to specialist rooms and teachers and, in fact, CDT will appear on their timetables. Even in these schools, however, most of the CDT activity will take place, not in the specialist room but in the classroom or year base and there is no real reason, other than teaching expertise, why pupils in junior or primary schools should not have similar opportunities.

In the DES Educational Pamphlet No. 57, 'Towards the Middle School', it is stated that:

> The drive to survive and therefore to master the environment may well be the basis both for curiosity and a desire for competence which are among the strongest incentives for learning. The key function of the middle school is to preserve and strengthen curiosity. Whether it succeeds depends on achieving a match between children's capacities and the problems they tackle.

(HMSO 1970)

Nowhere is this more true than in the age group under discussion. If pupils are to retain their enthusiasms they must be helped to achieve success and self-satisfaction at a time when they are beginning to become critical of their own abilities and impatient with failure. Their teachers must have very clearly defined aims and objectives and yet be flexible enough to be able to exploit the enthusiasms of the moment. They must have precise targets and yet be prepared to go off at a tangent if certain opportunities are presented. At the same time this is a most rewarding age range to teach and, in CDT, all sorts of activities begin to become possible.

Most of the CDT activity could be centred around 'Fields of Interest', and this approach is discussed at length in chapter four but, in short, the need is to select activities which will bridge subject barriers and which will give meaning and cohesion to the children's work. This approach is an extension of the topic work suggested for the 7 to 9 years age range and uses similar methods.

Because of the increasingly critical attitudes towards their own abilities and achievements shown by children in this age group it is important to provide adequate facilities for practical work, and this means finding space as well as appropriate tools and equipment if they are to find satisfaction with their efforts. In the average classroom this is difficult, although the trend towards falling rolls in the 1980s may put space at less of a premium in primary and middle schools. At the same time, practical work needs close and careful supervision for reasons of safety as well as to maintain and improve standards, and the class teacher cannot be in two places at once even if spare rooms do become available. This means that most practical work will continue to be done in the classroom with a full class of pupils and therefore space must be found by organizing the room in such a way that only a small group will be using tools and craft materials whilst the remainder are engaged in other activities. This has benefits, of course, as supervision becomes easier and the practical work can become part of the motivation process if it is interrelated with the other work under the umbrella of the 'field of interest'.

A possible field of interest for this age group could be 'wind movement' when the activity could encompass art work based on what is seen from the classroom windows or on topics such as 'a windy washday', trade winds and similar geographical content could be covered; the use of wind power for sailing ships and their historical development would be relevant; draughts and rising air from the central heating radiators might provide a scientific input, whilst the wealth of prose and poetry on the subject could be investigated as part of the English syllabus. In the area of CDT experiments with simple parachutes and paper gliders could lead to kite-making and flying, perhaps followed by the designing and making of simple flying machines powered by rubber bands. This last activity would need some research into the action of propellors and could be introduced by the making of simple paper 'twizzlers' made by cutting a square of strong paper diagonally from each corner almost to the centre and then bringing all four corners to the middle and securing to a dowel rod with a drawing pin. The introduction of washers to improve· the toy's bearing surfaces could be the subject of discussion and experiment and this would be the time to look at ships' propellers as well as those of aircraft. Children of this age range are ready to be involved in genuine problem-solving activity and the work listed above could be tackled in this way. Perhaps the problem could be to design and make a land yacht to be propelled by wafting it along with the draught produced by waving a square of cardboard and if more than one is made an exciting race could result.

Some of the ideas produced by the children will be beyond their immediate capabilities but the motivation could be used to encourage research into those aspects to form the basis of a written account or a talk to the rest of the class. This research and their investigations into their models would form the basis, together with sketches and cuttings, of their DESIGN BRIEF which constitutes an important part of the design process.

Where the more durable materials are used it is vitally important that correct tool handling and using skills and techniques are taught and all teachers are advised to acquire this expertise either by attending in-service courses to develop a mastery of basic tool skills and techniques, or by arranging for private tuition from a specialist colleague. Faulty tool skill cannot be corrected and it is very difficult to superimpose correct skills on a faulty foundation.

If technology is concerned with the use of knowledge, materials and resources to satisfy a human need then it follows that, ideally, the work done under other subject headings could be put to use in the problem-solving situation. Scientific knowledge is particularly important as a tool in the problem-solving process and therefore it is necessary to try to co-ordinate the work in CDT and Science. In fact, although the activity described is largely child centred, the amount of careful preparation needed exceeds that usually necessary for more didactic approaches.

From 11 to 13 years

Pupils in this age group can be preparing for transfer from middle to high schools at the age of 12 or 13, or they may already be in the secondary phase of education. In either case they are still middle years pupils with a need for a largely concrete approach to learning (Piaget's stage of concrete operational thought) although there is no doubt that their ability to think in more abstract and logical terms is widening provided that it can be related to something within their existing experience.

Problem-solving CDT activity can play a very important role in the educational programme of this age range as it enables the children to develop hypotheses using both concrete and abstract methods, to apply intuitive as well as deductive thought processes and to apply the results in order to check their reasoning in a concrete situation. In many cases these processes will be best carried out by experimenting, or 'playing', with the materials or equipment and by making simple models or prototypes in order to obtain a logical progression in the development of a solution.

Great care needs to be taken in selecting topics or problems to ensure that the pupils are working from a known or familiar starting point as an early 'breakthrough' can provide considerable motivation for the continuance of the activity and even the brightest pupil needs to refer to known or concrete examples when faced with fresh problems.

With pupils in middle schools it is still very easy for cross-curricular activity to take place as most classes will be organized in year bases and, although the amount of specialist teaching will have increased as the pupils progress up the school, there will normally still be a co-ordinated pattern to their activity. This means that work based on fields of interest could continue provided that there has been close liaison with the teachers who

I HEAR ·· I FORGET
I SEE ·· I REMEMBER
I DO ·· I UNDERSTAND

Figure 2.1 Aids to learning

have previously taught these groups in order to ensure that there is true progression and no unintentional repetition. The topics selected, of course, will be more challenging and will be investigated at a greater depth than has been possible with the younger age groups and there will be a need for a greater variety of resource material to be available to satisfy the pupils' needs. It is at this stage, perhaps, that the non-specialist teacher will be most apprehensive about personal deficiencies in the sphere of technological knowledge but really there is no cause for alarm as, almost invariably, there is no one correct answer or solution to the problem in hand and the need is to be able to point the pupil towards sources of knowledge or information whilst endeavouring to stay one jump ahead!

In the secondary phase, because of the pattern of greater specialization and the physical separation of specialist rooms and departments in many schools, it is more difficult to develop cross-curricular activity. However, there is no problem in arranging links between two or more subject areas, provided that the teachers concerned have time to talk to one another during their brief visits to the staffroom, and from this sort of beginning greater co-operation can follow. There would appear to be little doubt that cross-curricular co-operation leads to greater understanding of the relevance of subject matter than when subjects operate in complete isolation.

At this stage, too, it is very necessary to give all pupils the opportunity to spend more time in manipulating materials and in developing craft skills whilst engaged in problem-solving activity appropriate to their physical and mental development.

This activity must take place in properly organized and equipped workshops, and is as relevant for girls as it is for boys (see Figure 2.1).

Chapter Three

PROBLEM-SOLVING ACTIVITY

Technology and Science are closely linked and many teachers have come to Technology through activities they regard as practical Science. There are however distinct differences. Because Science is enquiry-led and discovery is for its own sake, the conclusions are drawn from the evidence and data and are as objective as possible. Technology, on the other hand, essentially involves meeting a need or solving a problem. The best solution will often involve a subjective judgement, and will be arrived at after taking a wide range of factors into account.

(National Curriculum Report: Science for ages 5 to 16: DES 1988)

THE PROCESS OF PROBLEM-SOLVING

Sometimes new ideas or approaches in education have to be regarded as passing fads or fashions, exploited by those who jump on the bandwagon but doomed to fade away in a relatively short space of time. External constraints coupled with grave suspicion of anything 'new' result in a resistance to change and a very conservative and traditional approach to educational method. The current interest in problem-solving activity, however, is not a new development as the method was strongly advocated by John Dewey (1859–1952) as part of his emphasis on child-centred learning and certainly most infant phase learning activity reflects his teaching and observations.

John Dewey was an American teacher, psychologist and

philosopher who believed very firmly that natural, dynamic learning situations could come about only as a result of the child's interaction with its environment and with other human contacts and that there was unlimited potential for this activity. He felt, too, that the emphasis on the presentation and absorption of facts, and didactic approaches generally, apparently unrelated to either the environment or to current need, had a stultifying effect in school on the learning processes and that, therefore, a change was needed from an emphasis on teaching to one based on learning situations. He was also determined to influence subject content in order to reduce the teachers' preoccupation with things relating to the past and to increase their awareness of the needs of the present and of the future.

If a pun can be allowed, certain of his beliefs were 'dewy-eyed' in that he appears to have assumed that all pupils would be naturally motivated by purposeful occupation with living situations but, assuming that the teacher has clearly identified appropriate aims and objectives for the work in hand, that the topic under investigation has been well chosen and that the pupils' lines of enquiry are guided and directed in an open-ended way, there is no doubt that the level of interest and satisfaction will be higher than in a chalk and talk situation, and the long-term educational and social benefits of the activity will be greater.

Dewey related the education process to the technological and social progress which man has made since his descent from the trees. Man's intelligence can really only be measured in terms of his ability to interact and react with his environment.

John Dewey's views, perhaps, were shared by his contemporary, Maria Montessori, who also believed that the teacher should play the part of a guide and organizer, helping the child to relate to the immediate environment and to use the child's interest in the topic under consideration to spark off spontaneous attention so that the learning process could be developed, guided and controlled.

Dewey was not in favour of progressive methods which left the child free to follow its own devices but he did believe very firmly that it was inquiry or, in other words, the process of thought, which enabled the child to learn about, and therefore adjust to, its environment and in order to control and direct this inquiry he expostulated three main procedural stages:

1 A preliminary review of the situation
2 Collection and analysis of appropriate material
3 Inquiry and experiment to test the truth and validity of the solution

These differ little from the stages recommended for investigatory activity as practised over many centuries.

Modern educational thought would suggest that purposeful activity in CDT is carried out in two main stages of development. The first stage is divergent in nature where the problem is investigated, evidence is collected and collated, possible solutions are proposed, and the net is cast as widely as possible, within the constraints set, to ensure that the second stage, when decisions will be taken, will be based on sound foundations.

The second stage is convergent in that it consists of narrowing down the options until finally a commitment is made and the selected solution is developed, constructed and evaluated.

In tabular form the process looks as follows:

Divergent stage	A	Investigate the situation
	B	Identify the problem to be solved
	C	Collect relevant information and investigate possible solutions
Convergent stage	D	Develop the selected solutions
	E	Construct or make the final version
	F	Evaluate and, if necessary, modify the finished product

Most CDT specialists would refer to this as the *design process* but design is an overworked and often misunderstood word meaning all things to all men and therefore the term *problem-solving process* is possibly a better label to use (see Figures 3.1, 3.2).

It is possibly worthwhile to spend a little time looking at this process more closely in order that the stages can be properly understood.

Stage A: Investigate the situation. This is where the pupil should be directed to find a need to design. By building con-

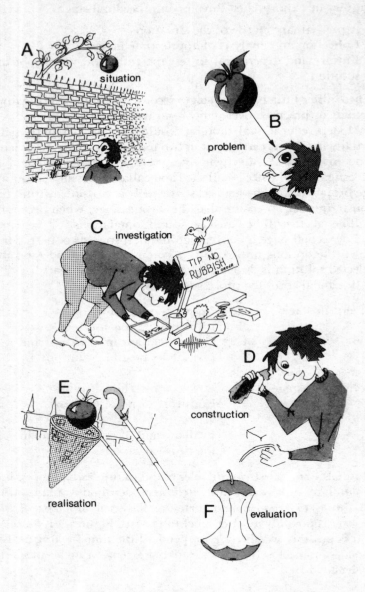

Figure 3.1 The problem-solving process

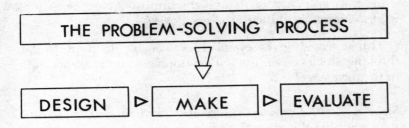

Figure 3.2 The design line

straints into the situation the activity can be kept within the ability level of the pupil to ensure that the outcome will have a good chance of success leading to a sense of achievement and personal satisfaction.

A possible situation which could be presented to an older junior school pupil for investigation could be that the tins which contain paper-clips and drawing-pins on the teacher's desk are frequently knocked off the desk top, spilling the contents all over the floor.

Stage B: Identify the problem to be solved. In this stage the pupil should be helped to draw up a statement of the need contained in the situation which has been presented. Many teachers refer to this statement as the 'brief' but this can be misleading as the folio of notes, sketches, information gleaned, working drawings and so on which are accumulated while the problem is being solved is called the 'design brief' and, under-standably, most pupils would be confused by the similarity of title. It is better called the 'problem statement'.

In order to control the teaching situation and to retain progression in the subject content it is sometimes advisable to include guidance as to suitable materials or processes in the statement, and from the situation already quoted in Stage A the problem statement could be:

Using the cardboard tubes and other materials provided, design and make a holder to contain drawing-pins and paper-clips for use on the teacher's desk. It must be possible to extract the pins and clips easily and the finished holder must look pleasant and be difficult to knock over.

This statement gives several clues to enable the pupil to start thinking along certain lines of development and leads naturally to the next stage.

Stage C: Collect relevant information and investigate possible solutions. At this stage there is a great temptation for children of all ages (as well as adults!) to seek out an existing solution and then to copy it – often using unsuitable materials or processes. As well as being most unsatisfactory in craft and aesthetic terms, copying of this type defeats the whole object of the exercise which is to apply intelligence to the problem in order to find the best possible solution within the constraints given.

A good approach at this stage is to require the pupils to start by listing as many of the factors to be considered as possible, and these could include dimensions, capacities, information about the materials which may be used, suggestions about suitable adhesives, accessibility to the parts of the artefact, appropriate surface finishes, and so on. The pupils would need to be able to handle the drawing pins and paper clips and to start to visualize the sort of containers which could possibly house them, and this activity would lead them naturally to the stage when they could start producing simple drawings in order to work towards a possible solution. They should be encouraged to produce a fairly wide range of drawings in order to explore the possibilities, with the drawings acting as captured thoughts. There is no reason at this stage why existing commercial products should not be examined and compared with the solutions produced and, if desirable, for modifications to be included.

With some problems drawing may be very difficult for younger children as, for example, when curved surfaces have to be shown. A very acceptable alternative is for ideas to be modelled in clay, capturing 3D ideas most effectively, or for shapes to be cut out from paper with scissors and used instead of drawings.

On completion of the exploratory processes the convergent stage begins and this is where decisions are taken and the pupil is committed to a main line of development. Modifications are still possible, of course, and in fact they are often desirable as they will arise out of actual experiences as the work progresses, but the overall plan of campaign should be drawn up before practical work commences.

Stage D: Develop the selected solution.   When the decision has been taken, with the teacher's guidance, as to which of the proposed solutions is the most suitable, a neat dimensioned drawing or drawings should be produced to act as a source of information. This drawing should be simple but clear and should contain enough information to enable the pupil to work from it with the minimum of teacher assistance.

Stage E: Make the selected version.   It is often a good idea for the pupil to make a quick mock-up or prototype before commencing work on the final model, perhaps using paper or thin card held together with gumstrip. If new tool skills or processes are to be introduced these must be taught and reinforced as the need arises. At no time should a pupil be left to find out how to use a tool for himself as skills must be correctly taught and supervised at all stages of education. During the construction stage very high standards of finish should be demanded by the teacher.

Stage F: Evaluate and, if necessary, modify the finished product.   Educationally this is one of the most important stages of the process as the development of an attitude of critical awareness is an essential part of the problem-solving process. The pupil must ask himself 'does it satisfy the need?' or, in other words, 'does it do the job for which it was intended?'. It is highly unlikely that the finished product will satisfy the need in every respect but so long as the evaluation process highlights the modifications which may improve it then that will suffice. If the pupil is sufficiently motivated to want to make an improved version then this should be encouraged as long as it will not hamper or delay further progression in the subject (see Figure 3.3).

Although problem-solving activity has been presented here as if it is a linear process this is really an oversimplification as at any point it may be necessary to return to an earlier stage in order

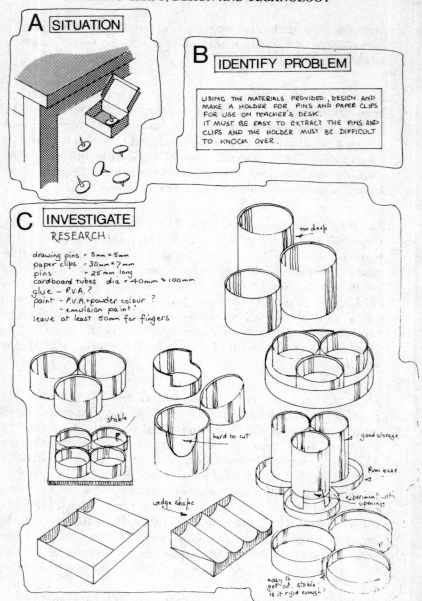

A SITUATION

B IDENTIFY PROBLEM

USING THE MATERIALS PROVIDED, DESIGN AND
MAKE A HOLDER FOR PINS AND PAPER CLIPS
FOR USE ON TEACHER'S DESK.
IT MUST BE EASY TO EXTRACT THE PINS AND
CLIPS AND THE HOLDER MUST BE DIFFICULT
TO KNOCK OVER.

C INVESTIGATE

RESEARCH:

drawing pins = 5mm × 5mm
paper clips = 30mm × 7mm
pins = 25mm long
cardboard tubes dia = 40mm & 100mm
glue – P.V.A.?
paint – P.V.A.+powder colour?
– emulsion paint!
leave at least 50mm for fingers

too deep

stable

hard to cut

wedge shape

good storage

firm base

experiment with openings

easy to get at, stable,
is it rigid enough?

Figure 3.3 An example of problem-solving

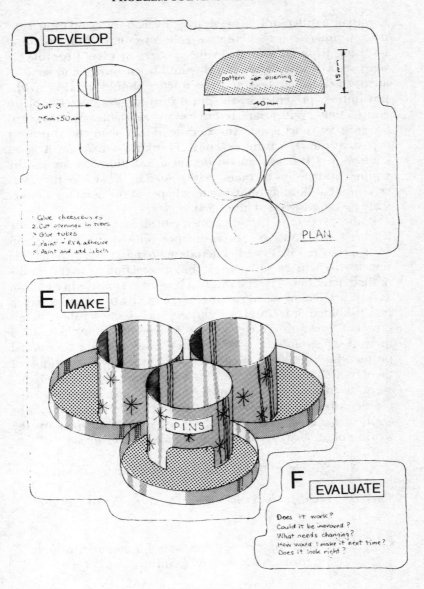

D DEVELOP

Cut 3
75mm × 50mm

pattern for opening

15mm

40 mm

1. Glue cheeseboxes
2. Cut openings in tubes
3. Glue tubes
4. Paint + P.V.A. adhesive
5. Paint and add labels

PLAN

E MAKE

PINS

F EVALUATE

Does it work?
Could it be improved?
What needs changing?
How would I make it next time?
Does it look right?

to pursue a different line of enquiry. Even at the evaluation stage new approaches or alternative solutions may come to mind resulting in modifications to the product or even a completely fresh start. Normally the pupil would be encouraged to see it as an open-ended process and the teacher should be alert to the possibilities presented by the pupil's investigations to be able to ensure that appropriate resources are available and that real progression is achieved. An example of this open-ended process occurs in the DES film 'Practical Thinking' where an 'A' level student in CDT sets out to design a kaleidoscope for use by young children but becomes so interested in the shapes that he is creating that he ends up using the shapes as the basis for a large-scale rocking toy (see Figure 3.4).

If the process is to be as open-ended as possible, then the teacher's role must be to ensure not only that adequate and appropriate resources and materials are to hand, but also that the pupils will achieve both success and satisfaction from the results of their activities. The end product may be decorative in function as, say, an item of wood carving; it could be utilitarian as, say, a pencil-holder; or it could be technological as, say, a rubber-band-powered wheeled vehicle, or it could be any combination of these. For example, a wooden paper knife serves a functional purpose for only a very small proportion of time and for most of the day it should be regarded for its sculptural qualities. However, no matter what form the end product may take, the problem-solving process is equally applicable. Indeed it can be applied to almost any aspect of school activity and perhaps the writing of an essay could be seen as one such example:

Stage A: *Situation*: this is where the motivation is found

Stage B: *Identify the problem*: this leads to the title of the essay

Stage C: *Collect information, and so on*: notes and jottings are made as ideas come to mind

Stage D: *Development*: the ideas are sorted out to be used in paragraphs or sections

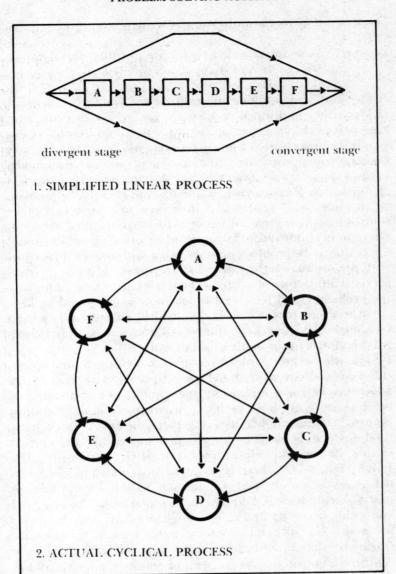

Figure 3.4 Two views of the design process

Stage E: *Construction*: the essay is written

Stage F: *Evaluation*: perhaps parts are rewritten and the final essay is then read aloud or displayed.

If the pupils are working in groups quite young children can tackle fairly complicated problems as each child, or better still, each pair of children, can tackle part of the overall problem. For example, it may be that a group of 7-year-old boys and girls have been presented with the situation that their sand tray model needs a device to lift water from the pool into the irrigation channels. Simple research will lead them to identify the problem in that they will realize that they need to design and make something which eventually, they will learn, is called a shaduf. One pair of children could investigate the principles of levers and apply this to the problem, another pair could concern themselves with the counterweight whilst a third pair could work on ideas for the manufacture of a suitable bucket-like container. This way the problem could be solved much more quickly and be kept down to single problem elements enabling short-term targets to be achieved and, of course, this sort of teamwork simply reflects what happens in the adult world of work.

Similarly, a group of 10-year-olds designing a simple anemometer could work separately on the problems associated with attaching the cups, supporting the spindle, providing suitable bearing surfaces and, perhaps, inventing simple counting devices, bringing all their ideas together to be incorporated in the final model to be made by the group. Some teachers are worried in case they are too helpful towards their charges during CDT activity but, really, there is no point in requiring children to rediscover the wheel! Surely the aim should be to incorporate up-to-date technological knowledge as far as it is suitable and applicable to the age and ability range and to require the pupils to modify, adapt and improve technological components. During the developmental stage principles and feasibility can be confirmed by using string, paper and Sellotape approaches but then teacher intervention is very necessary in order that the construction stage can be carried out with a high level of craftsmanship using suitable materials and appropriate skills and techniques which, of course, must be correctly demonstrated and supervised by the teacher if acceptable standards of finish are to be achieved.

During the evaluation stage it is sometimes useful for the teacher to be able to suggest objective questions rather than for the children to apply purely subjective judgements, and the questions which a craftsman could use in this situation would include the following.

- Is the end product well made?
- Have the correct tools, processes and materials been used in its construction?
- Does it do its job properly? (that is, is it functional?)
- Is it appropriate for its environment?

If a negative answer is received to any of these questions then appropriate modifications would appear to be desirable. Incidentally, if the adult consumer asked himself these questions every time he considered buying some durable domestic product then there would be a greatly reduced market for many of the shoddy items at present to be found in shops and supermarkets and it would not be long before certain manufacturers would be forced either to improve their products or to go into liquidation.

As well as the finished model the problem-solving process should result in a design brief being produced containing the notes and drawings produced by the pupil together with information on difficulties, trials and successes as well as details of sources of information used in arriving at the solution. These design briefs can then become part of the pupil resource material available subsequently to anyone else tackling a similar problem. By this means progressively higher standards can be achieved as pupils benefit from, and improve upon, the results of earlier work.

If this process is to lead to good motivation, with all pupils benefiting according to their age and ability levels, then the responsibility for selecting appropriate situations and problems rests heavily with the teacher. It is important that each child achieves success in order to gain satisfaction from its work, but this must be achieved whilst progression is being attained with the child moving forward to new ground and fresh challenges. Actually, this is not as formidable as it sounds because the only real difference between the problem-solving and the traditional approach to CDT is that in the former the child is given the situation to investigate rather than being given a predetermined solution to copy. The actual content of the scheme of work using either method could be very similar but with problem-solving activity the prime aim is to involve the child as much as possible in the educational processes rather than in pure imitation and rote memory.

Surely, then, this means that the teacher must have hundreds of problems ready and to hand to cater for the needs of the mixed ability class? Fortunately, in practical terms, this is not the case as most situations are capable of investigation by a wide age and ability range. To take an example well within the average primary school teacher's present experience, the infant, the junior and the 'A' level pupil could all be working on puppetry. Each pupil could be stretched whilst producing an excellent puppet for the age level and yet the practical and intellectual involvement and the educational achievement would vary enormously with the different stages of schooling. So it is with many of the problem situations suitable for CDT work, when any one problem can be used with a wide age and ability range provided that the teacher incorporates suitable constraints limiting the materials available or the techniques and skills needed to enable the pupil to reach the evaluation stage successfully and to achieve high standards of craft skill and 'finish'.

Many Local Education Authorities run annual Science and Technology competitions to encourage technological development in the schools and one such Authority, Staffordshire, has run a successful competition sponsored by the British Association for the Advancement of Science each year since 1977. The competition is open to all ages and it seeks to encourage pupils to grapple with a problem outside the traditional paths of the curriculum but which reflects situations facing scientists and technologists in industry and research. Each year the competition attracts large numbers of entries from boys and girls aged from 5 to 18 years but the interesting fact is that by dividing the judging into three age groups – (1) under 11 years – up to top juniors; (2) 11 to 14 years – 3rd and 4th year middle and 1st, 2nd and 3rd years secondary; and (3) over 14 years – 4th, 5th, 6th years secondary, or under 19 years, full time, in a college of Further Education – it has been possible to set one problem only each year, with slightly different objectives for each age group, and the results achieved by the younger pupils are often more exciting and much more successful than the more sophisticated entries of the older students. Because of time and accommodation constraints the competition allows three entries only from each age group in each school but most schools run their own eliminating competitions and several schools run subsequent local interschool events so that very large numbers of pupils are ultimately involved in the annual event. Each year the best entries are from schools where access to both scientific and technical help has been possible.

EXAMPLES OF SCIENCE AND TECHNOLOGY COMPETITION PROBLEMS
1 Rubber-powered, wheeled-model vehicle

The wheeled vehicle must be rubber-powered and self-contained. It is to run on a circular test track with an internal radius of 1m and a width of 250mm. The vehicle will be tethered to a ball race on a post in the centre of the track, and during the run at least one wheel of the vehicle must remain in contact with the test track at all times.

Entries will be judged on the total number of circuits made of the track, the best of two runs to count. Each entry must be accompanied by a design brief outlining the development of the vehicle.

Most of the entries in the early and middle years age groups consisted of models with the power going directly to the wheels but, of course, these had problems in controlling the release of energy to drive the wheels consistently over as long a period as possible without the wheels slipping or spinning. The most successful were those using some form of gearing between the power source and the driven wheels. Entries ranged from wheeled, powder-colour-painted soap boxes to carefully engineered entries based on bicycle wheels.

The entry details were accompanied by a dimensioned drawing of the track so that the schools could simulate the test conditions.

2 Model glider or heavier than air machine

Pupils are required to design and make a glider or similar heavier than air machine which may be hand or mechanically launched. The machine must not exceed 150g in weight and no mechanical, gaseous or other aids may be used to support the machine in flight. The entry must be accompanied by a design brief and the launcher can form part of the design brief.

No commercially produced kits or aerodynamic parts may be used. The winning entry for each age group will be the model which remains in the air for the longest time from point of launch to point of contact with the ground. Entries will be judged on the better of two flights.

Most of the entries for this competition consisted of some form of glider, the simplest being hand launched and made from polystyrene ceiling tiles. Others used model aircraft construction techniques with rubber-powered or compressed air launching devices. One of the section winners produced a parachute launched from the end of a specially adapted fishing rod which had a release mechanism fitted and the competitor carried out the launching from the top of a tall pair of steps so that the parachute commenced its fall from a point some 10 metres or so from the ground. This was hardly what the organizers had in mind when setting the problem but it was a good example of ingenuity within the conditions of entry.

3 *Model boat or water-borne vessel*

A toy manufacturer wishes to produce a powered vessel to move on a water surface and pupils are invited to design and make a prototype model or models which will be judged on speed over a measured distance in a test tank to be provided by the organizers. The vessel may be powered by any means but the power source, mechanism and controls must be contained within the vessel. The model must remain in contact with the surface of the water for the whole of the test run.

The test tank will consist of a 10m length of half-round plastic guttering 150mm diameter, 75mm deep, which will be filled with water to a depth of 50–60mm.

The models will be timed over the following distances:

Under 11 years	4 metres
11–14 years	6 metres
15–18 years	8 metres

The test runs will be timed electronically and models must be made so that they have some part at least 20mm above the level of the guttering to trigger the timing mechanism. Models will be timed on the better of two runs and they must complete the allocated distance to qualify.

Each entry must be accompanied by a design brief showing the development and modifications made from the initial ideas to the completed model.

Here, entries ranged from rubber-powered models, vessels using model aircraft engines and others jet-powered by using Sparklets bulbs. One middle school entered models powered by home-made rockets, but in view of the fact that they succeeded in setting off the school's smoke alarm system (resulting in the school being evacuated while the competitors carried on blithely unaware of the drama!) it was decided to ban all forms of explosive device from all future competitions.

The biggest problem faced by competitors in this competition was to obtain maximum speed whilst retaining contact with the water surface. The best entries aquaplaned while slower ones remained deeper in the water and others were disqualified because they became airborne.

4 *Target competitions*

Pupils are required to design and make a machine to project a standard table tennis ball at a target (drawing provided). The target comprises a board 1330mm high by 600mm wide with four holes on the centre line of the board sized 380mm, 300mm, 230mm and 150mm diameter respectively. The target is supported on two blocks allowing it to be inclined to the horizontal at 45°, 60° and 90°. Balls projected through the holes will score as follows:

Upper hole	150mm	25 points
Second hole	230mm	5 points
Third hole	300mm	3 points
Lowest hole	380mm	1 point

Each entry will be allowed three shots at the target and the total points scored will count towards the score for that entry. The minimum distance between the point where the ball leaves the machine and the base of the target, and the angle at which the target is erected will differ for each age group as follows:

Under 11 years	2 metres	vertical
11–14 years	4 metres	60° to the horizontal
15–18 years	6 metres	45° to the horizontal

No connection will be allowed between the projecting device and the target, explosive devices are forbidden and direct propulsion by human effort will not be allowed.

Each entry must have a supporting design brief which records the investigation, development and evaluation of the device.

The winning entry in each age group will be determined by the largest total score obtained by combining the target scores with the score for the design brief.

This competition gave scope for some varied and ingenious solutions. Most pupils used some sort of variation on the Roman catapult but one school, unsuccessfully, produced a huge launcher made from tubular steel barriers, the projectile being launched by swinging a large bass broom from a central pivot to produce a heavy-handed golf swing!

Among the middle and early years entries were devices powered by vacuum cleaners, mechanical launchers made from Meccano and also rubber-powered mortars. The standards of accuracy in this competition were very high and the final grading depended on the marks awarded for the design briefs.

5 Bridge-climbing vehicle

Pupils are required to design and make a self-propelled vehicle to negotiate the course and stop on a predetermined target. From the starting point the vehicle is to travel along a polished wooden floor to the bridge which will be 1200mm long, 300mm wide and rising to a crest based on a 300mm radius curve. The vehicle must climb the hill and descend on the

other side progressing to a fan-shaped target area marked out
in 300mm bands (drawing provided) with scores of 5, 10, 25, 10
and 5 respectively. The target area will be marked out on the
polished floor with masking tape.

The starting point will be as follows:

Under 11 years	600mm from bridge
11–14 years	1200mm from bridge
15 years and over	1800mm from bridge

The vehicle must be self-propelled and guided. No external
power source or control will be allowed nor will any contact
between the vehicle and the target area. Propulsion by direct
human effort or by explosive devices will not be allowed.

Each entry must be supported by a design brief and the
winning entry in each age group will be determined by the
highest score obtained by combining the target scores with the
scores for the design brief.

Entries for this competition ranged from rubber-powered
squeezy bottles to machines incorporating mercury-level switches
and on-board computers. The simplest machine consisted of two
large squeezy containers with the bottoms cut off and the two
joined together to make one long cylinder. A large rubber band
was fixed through the middle and attached to a length of cane on
the end of which was a wooden ball. The cane was revolved to
twist the rubber band to a predetermined tension and then the
device crawled along the route. Each time it passed beyond the
target area and then stopped and, using the remaining stored
energy, reversed onto the maximum score three times out of
three!

This was a complex problem as the devices had to be able to
climb a hill without wheel spin, descend on the other side
without slipping while still travelling in a straight line and then
stop at a predetermined distance from the starting point.

To be more effective this problem needs a time penalty as some
devices were painfully slow and reduced the excitement so
necessary for a lively competition.

6 *Cable climber*

Pupils are required to design and make a device which will travel along a wire and release a ball bearing or glass marble over a target. The wire to be used is 1.5mm diameter stranded Bowden cable under tension at a height of 1.5m from the floor. The distance between the vertical supports is 4m and it is expected that a depression of 50mm will take place at the centre of the wire when subjected to a mass of 0.5kg, the maximum mass of the device to be allowed. The device will be required to travel between two measured points 3m apart in a time of less than one minute. It must carry a ball bearing or glass marble not exceeding 9mm diameter, which is to be released on to a target area situated below the wire. A 500mm run-in will be allowed at the start and there will be a similar clearance at the end of the run. The maximum score on the target will be 2m from the start point.

The target will consist of a sand tray 1m long by 300mm wide marked out in sections scoring 10, 20, 30, 40, 30, 20, 10 points respectively. The 'bull's eye' will score 50 points and will consist of a bicycle bell top suspended exactly in the centre of the 40 point section. The measured points of 500mm from the start and finish must not be crossed before release and the device must not extend more than 300mm below the wire. The device is to be self-contained and no external control will be allowed.

Each entry will be allowed three runs with each score being totalled for the entry. No trial runs will be allowed. Methods of propulsion which in the opinion of the judges are considered dangerous will not be permitted. Each entry must be accompanied by a design brief which records the development of the device and its specification and this design brief must have been prepared totally by the pupils.

Standards for this competition were high and a full range of energy-producing devices was used including clockwork and electric motors, internal-combustion model aero engines, rubber bands, balloon power and squeezy bottle rockets. The competitors using heavier power sources tended to compensate for this by making the remainder of the device from polystyrene ceiling tiles.

Problems encountered by the pupils included preventing the

device from swinging from side to side as it reached the middle of the cable, and difficulties in retaining power and traction as it climbed up the incline on the second half of the journey.

7 *Wheeled projector*

Pupils are required to design and make a rubber-propelled vehicle which will travel between two points and in so doing project one or more table tennis balls (depending on age group) on to a horizontal target situated at right angles to the direction of travel of the vehicle. External control of the vehicle is not permitted.

The vehicle will travel along the floor for a distance of 6m and a standing start is not necessary, but from the crossing of the official start line a maximum of 60 seconds is allowed to complete the 6m track. The number of unused seconds from this time counts in full towards the final score, for example, a vehicle which takes 35 seconds to complete the 6m run has 25 points added to its target score. Any vehicle taking longer than 60 seconds to complete the run will be disqualified.

The target is situated 1m from the run and consists of a triangle with a base of 2440mm and a centre length of 2440mm marked out in horizontal sections. The base section scores 25 points, the centre sections score 50 and 75 points and the apex section scores 100 points. The score, as decided by the judges, will be that part of the target with which the table tennis ball makes initial impact:

Under 11 years – fire ONE table tennis ball during the run
11–14 years – fire TWO table tennis balls during the run
Over 14 years – fire THREE table tennis balls during the run

Each vehicle is required to make three runs with each score counting towards the total (plus time points and the score for the design brief). Practice runs are not permitted and the vehicle must not cross the line towards the target.

Each entry is to be accompanied by a design brief concisely recording the development of the vehicle and its specification. A maximum of 50 points will be awarded for the design brief.

Pupils had obviously benefited from earlier competitions and the ideas used tended to be amalgamations of earlier wheeled vehicles and launching devices. The time penalty led to an exciting competition and those entries which put all their eggs into one basket by firing their two or three table tennis balls simultaneously tended to score less highly than those who attempted the harder solution of firing at the target separately.

One very ingenious entry which scored well consisted of a wheeled board powered by a rubber-band-driven propeller and the projectile launcher appeared to have been inspired by the Mousetrap game. A glass marble was placed in the top of a contorted arrangement of domestic plastic tubing and joints of the sort used by plumbers for cold water installations and on falling from the bottom of the tubing it landed on a platform which triggered off a catapult to send the table tennis ball on its way! Many hours must have been spent by the design team to get the timing right on this one!

It is encouraging to note that the number of entries from primary and middle school pupils invariably exceeds that from secondary-aged children, and that girls are generally well represented. Certainly there is more evidence of ingenuity and originality of thought from the primary pupil competitors, and the enthusiasm and excitement of the competition day is something not to be missed.

More important, though, is the fact that the first school pupil and the 'A' level candidate can, in fact, tackle a common problem, and although the solutions will differ enormously in complexity and technological and scientific content, it is possible for all children to achieve a high standard of craftsmanship and to be fully extended in educational terms, gaining enormous satisfaction from the sense of achievement which the finished model brings, whether or not it is a major prize winner. To help with this attitude all entrants for the competitions in Staffordshire receive a signed certificate of entry and, of course, the winners receive certificates together with trophies kindly donated by supporters from industry.

Specimen problem-solving situations

1 *Situation*: There are times when the tins containing drawing-pins and paper-clips are knocked off the teacher's desk, spilling the contents all over the floor.

Problem statement: Using the cardboard tubes and other materials provided, design and make a holder to contain drawing-pins and paper-clips for use on the teacher's desk. It must be possible to extract the pins and clips easily and the finished holder must look pleasant and be difficult to knock over.

2 *Situation*: Because of gravity, unsupported items invariably fall to the ground.
Problem statement: Modify the sheet of A4 paper which you have been given so that it stays in the air for as long as possible

3 *Situation*: New pupils to your school could have difficulty in finding their way around the building and visitors could need help in finding the Secretary's office.
Problem statement: Design a system of maps and/or colour-coded signs which will guide a visitor round your school. Study the London Underground map and similar guides to help with your ideas.

4 *Situation*: Notes telling the milkman how many pints are needed can blow away or be made illegible by rain.
Problem statement: Design and make a device to let the milkman know how many pints to leave. Your solution must be easy to use and very clear to read.

5 *Situation*: Children love playing with toys, especially if they 'do' something.
Problem statement: Design and make a balancing toy to be part of a child's circus. The toy should be made so that it will balance on a taut length of thin string. The finished toy must be colourful and safe for a young child to use.

6 *Situation*: You will have come across the term 'tessera' in mathematics. It is fun to discover shapes which will fit together to make a mosaic.
Problem statement: Design and make a shape so that several pieces of the same shape will fit together to make a toy which will float on water.

7 *Situation*: Paperback books are easily damaged or lost if they are left around the house.
Problem statement: Design and make a holder which will con-

tain ten paperback books in such a way that their titles are visible and so that they are easy to take out and replace.

8 *Situation*: Signalling is fun if you know the Morse code or can read semaphore.
Problem statement: Design a circuit which will operate off a torch battery and which can be used to flash code messages to a partner. Your device must be neatly housed and self-contained in a tray or box.

9 *Situation*: A model is always much more fun if it 'works'.
Problem statement: Design and make a model car, lorry or bus by sticking together and painting some small cardboard packets (custard powder, dried fruit and tea packets are about the right size). Find out how to make a 'motor' from a cotton reel, rubber band and a pencil or piece of dowel rod. Cut the bottom out of your vehicle and use the 'motor' to make it go along. You could hold competitions with your friends to see which goes the fastest or the longest distance or you could use them to bring life to a model village.

10 *Situation*: Perhaps you have a collection of Lego at home or at school and all the bits are jumbled up in a large box making it difficult to find the parts you want.
Problem statement: Design and make a storage box to hold Lego pieces so that the different parts can be found easily. The materials available are cardboard, gumstrip, paper, glue and a piece of hardboard measuring 250mm by 200mm.

11 *Situation*: Without knowing the language it is difficult to communicate with someone from another country.
Problem statement: Without using words, design a poster or a device which could be used to show a child from another country how to play hopscotch.

12 *Situation*: When a recipe book is being used to help prepare a meal it is difficult for the 'cook' if the page turns over in a draught just as the cook's hands are covered in flour.
Problem statement: Find a way of holding a recipe book open at the correct page and so that the page remains clean.

13 *Situation*: Most people make a collection at some time in

their lives. Perhaps you have a collection of sea shells or foreign coins or fossils or interesting pebbles.

Problem statement: Design and make a simple box or storage tray to keep your collection tidy and to make it easy to display.

14 *Situation*: Birds are a nuisance when they scratch out Dad's freshly sown vegetable seeds.

Problem statement: Design and make a device which will frighten birds away from the vegetable plot. Perhaps your model could be made to move by the wind.

15 *Situation*: Empty coffee jars make very good and attractive storage containers for the kitchen.

Problem statement: Design and make a set of six labels to identify the contents of the jars when in use, either for tea, sugar, salt, flour (plain and self-raising) and custard powder or for six types of spice. The labels could be stuck on with PVA adhesive.

16 *Situation*: Look at a brick wall in your school. You will see that the bricks are overlapped in rows . . . this is called a BOND. Why are walls built in this way?

Problem statement: Using wooden or plastic bricks, build a wall in the same arrangement as the school wall. See how many rows you can build before it falls over.

17 *Situation:* Oxo and Bovril meat stock cubes are useful when cooking.

Problem statement: Design a holder for use in the kitchen to store and dispense beef and chicken stock cubes. The cubes should be dispensed one at a time and the holder should be easy to refill.

18 *Situation*: Squeezy bottles filled with pebbles, peas or rice make an interesting sound when shaken.

Problem statement: Make three differently sounding musical instruments based on squeezy washing up liquid plastic bottles. When painting the containers how can you get the paint to stay on the rather 'greasy' surface?

19 *Situation:* The Greeks and Romans used war machines which would hurl a lump of rock at the enemy's defences. Look at pictures of these machines to find out how they worked.

Problem statement: Using Balsa wood, or folded card, design and make a war machine which will project a table-tennis ball accurately for a distance of at least 1 metre.

20 *Situation*: The housewife uses a clock, a pinger or an egg-timer when boiling an egg for breakfast. Imagine that you are marooned on a desert island where there are plenty of sea birds' eggs but no clocks or watches.
Problem statement: Using only natural materials, or items which could be washed up by the tide, design and make a timing device which will warn you when approximately four minutes have elapsed. If possible the device should also make a noise when the time is up.

21 *Situation*: Dominoes make a popular game but in order to play children must be old enough to count.
Problem statement: Design and make a game which can be played like dominoes by matching pairs of cards or blocks but which does not rely on the player being able to count.

22 *Situation*: Blind people normally have a very good sense of touch.
Problem statement: Design and make a game which could be enjoyed by two blind players. Test your solution by playing with it blindfolded.

23 *Situation*: Making a vehicle which will climb up a low hill and then descend smoothly presents all sorts of problems.
Problem statement: Design and make a rubber-band-powered vehicle which will climb up a hill and then proceed under control down the other side.

24 *Situation*: A static model windmill can be made fairly easily from empty cardboard containers glued together and painted but it would be much more interesting if the sails could be made to turn.
Problem statement: Design and make a windmill with sails that will revolve when you turn a handle at the back. Perhaps you could use Lego gear wheels from a Technical Functions set or you could make your own by gluing strips of corrugated card around empty cheese-segment cartons and spice drums.

25 *Situation*: Windsurfing and land yachts provide extra thrills when holidaying at the seaside.

Problem statement: Using Lego and paper sails, design and make land yachts which can be blown along by wafting them with a piece of card. Race your models and find out why one may be faster than another.

26 *Situation*: The need to span a gap occurs frequently in the environment.
Problem statement: Using only the three sheets of A4 paper provided, design and make a structure which will allow a model car to be rolled along it when it is supported at 25mm from each end on housebricks.

27 *Situation*: By careful packaging it is possible to protect very fragile objects whilst in transit or storage.
Problem statement: Use the kitchen foil provided to make a packaging for an egg, which will enable it to be dropped from a height of 200mm without breaking.

28 *Situation*: Tall structures are often needed in today's crowded environment.
Problem statement: Using only the drinking straws and pipe cleaners provided, design and make the tallest structure you are capable of building.

29 *Situation*: Dragster racing is a popular sport using full-size or model vehicles.
Problem statement: Design and make a model dragster from card using wheels and axles which you will make yourself. The model should be powered by an elastic band. The dragster should travel as fast as possible over a distance of 4 metres.

30 *Situation*: Certain scales are available to weigh foodstuffs and others are designed for bathroom use. In industry, however, specialist scales are sometimes needed to weigh specific items.
Problem statement: Design and make a device which will weigh liquids in half-litre lots. Accurate weights will be available to enable you to calibrate your solution. The container for the liquid must be demountable to enable the liquids to be poured back into other containers.

It will be realized that most of the examples given are capable of being adapted for use with a wide age and ability range and many of them do not require specialist workshop facilities or tools other than those found in most classrooms. In the upper two years of the middle school and in the secondary school, however, many of the situations and problems set would require specialist knowledge and equipment and would be concerned with the design and manufacture of artefacts or systems for home or school use.

Examples of the sort of problems which could be set requiring the use of more durable materials might include:

1 *Situation*: A nail on the back of a door is hardly the best way to store a jacket when it is not being worn.
   Problem statement: Design and make a device which will hold a jacket by the loop when it is not being worn.

2 *Situation*: It is dangerous to ride a bicycle without effective front and rear lights but batteries run down giving a poorer and poorer light.
   Problem statement: Design and make a holder which can be carried on a bicycle to hold a spare set of batteries. The holder must be easy to open and be able to keep the batteries dry.

3 *Situation*: It is common practice nowadays to have a light meal whilst sitting watching television.
   Problem statement: Design and make a device which will safely hold a plate and a cup or beaker so that a snack can be enjoyed while sitting in an easy chair watching television.

4 *Situation*: Most recipes call for the use of salt, pepper or herbs.
   Problem statement: Design and make a rack which will hold and display a range of kitchen herbs and spices. Remember to start by examining and measuring the appropriate jars or containers.

5 *Situation*: Elderly people and others suffering from rheumatism or arthritis find it difficult to open screw-top jars or bottles.
   Problem statement: Design and make a device which will make it easy to remove screw caps and lids.

6 *Situation*: Outdoor activities take place in all weathers and it is often difficult to find a spot dry enough to sit down.
   Problem statement: Design and make a collapsible seat for use when watching sport or when fishing. The seat should fit

into a canvas bag measuring no more than 300mm by 350mm by 100mm thick when folded.

7 *Situation*: Sometimes doors need to be propped open for short periods.
Problem statement: Design and make a device which can be operated and released by the foot and which will hold a door open when needed.

8 *Situation*: Making scale models or assembling plastic kits necessitates a small but essential kit of tools including knives, needle files, razor saw, scissors and tweezers.
Problem statement: Design and make an attractive box which will house a model-maker's tools and equipment enabling him/her to remove them easily for use.

9 *Situation*: Most families acquire a collection of audio or video tapes.
Problem statement: Using either hardwood, or acrylic resin sheet, design and make a rack for the storage and display of your set of tapes. Provision should be made for the addition of extra tapes as your collection grows.

10 *Situation*: It is dangerous to balance on a ladder to pick apples hanging from the ends of branches and if they are knocked off the branch they will bruise and will not keep.
Problem statement: Design and make a device which can be used from the ground to harvest ripe apples from the branches of a tree.

In selecting situations and problems for the pupils to solve it must be remembered that there should be a progression in the acquisition of both knowledge and skill and it will often be necessary to build in constraints in the use of materials or to specify certain types of construction to enable this to be achieved.

Chapter Four

TECHNOLOGICAL AWARENESS

As children develop their technological capabilities, they need to develop their awareness, capability and understanding, all of which are closely related.

At the primary stage, children learn about the world around them mainly through first-hand experience and with the help of teachers develop important skills, concepts and attitudes. They do not see the boundaries between one form of knowledge and another.

(National Curriculum Report: Science for ages 5 to 16: DES 1988)

TECHNOLOGY ACROSS THE CURRICULUM

In a few secondary schools Technology is being introduced as a new, additional subject complementing work in design and in physical science but remaining separate from them. In the majority of middle and secondary schools, however, technology is to be found as a component in a broadly-based CDT programme rather than as a subject in its own right, but what is certainly lacking in many schools is a positive approach to the establishment of cross-curricular activity leading to technological awareness, as distinct from technological ability, on the part of both teachers and pupils, and this aspect is as important to the early and middle years as it is to the secondary phase.

What is needed in the early and middle years phases is not a new subject but the recognition that much of the present subject content across the curriculum already has technological relevance and that this should be exploited to make existing courses more effective and more relevant to present and future needs. A development of this sort would inevitably include work in craft and design giving the concrete aspect so necessary in the work with younger children, and could provide the common ground needed to co-ordinate the curriculum.

Any topic which relates to man's domestic, industrial or social development, or to his interaction with the environment, must, by its very nature, have technological connections and these should be identified by the teacher and given due prominence. What is being discussed here is an extension to existing subject matter or, in some cases, a change of emphasis to increase the children's technological awareness and to relate the lesson content to human needs and resources.

Very few teachers, other than CDT specialists, will have had an opportunity to attend courses of training relating to the acquisition of technological skills or awareness and yet it is essential that all teachers should be able to exploit technological concepts and possibilities in their existing work in the classroom if their teaching is to be really relevant to present-day needs. No great technological knowledge or skill is needed for this change of emphasis but many teachers would benefit from help and advice offered by colleagues with practical and social abilities and interests or, better still, by co-operating with other teachers to introduce broadly-based 'fields of interest' aimed at bridging traditional subject barriers but providing opportunities for inventive, imaginative, creative work whilst reinforcing the basic skills of literacy and numeracy and in improving communication skills. If it is accepted that Piaget's 'concrete stage' of development is applicable to the majority of pupils throughout most of the 5–13 age range, it follows that an emphasis on practical work, particularly designing and model-making, supporting a broadly-based approach to topic work, could be the most effective way of establishing technological concepts whilst increasing basic skills. If the work is carried out by small groups of pupils then there will also be added opportunities for the development of communication skills between members of the group as well as for individuals to rationalize their own thoughts.

Part of this process should be to create opportunities for pupils to make models and other things that really 'work' or, at least, demonstrate a principal and that they should always be encouraged to evaluate the end product, or prototype, as an important part of the educational process.

In view of the fact that children in the 5–13 age range could be in any one of a wide range of schools offering very different facilities and teaching skills, it is very important to try to give them equality of opportunity no matter what the learning environment may be. It is therefore vital that all girls and boys should be given genuine CDT experience irrespective of the type of school in which they are receiving their education as it is during these early and middle years stages that attitudes are developed and consolidated. The three components of craft, design and technology suggest that the work should have both a technological and an aesthetic content and, therefore, there should be an introduction to the application of scientific, mechanical and electrical knowledge as well as an encouragement of the acquisition of an extended vocabulary to enable aesthetic consideration and functional evaluation to take place within the existing, but broadened, curriculum.

Technological extension of existing subject matter

As an example of existing subject matter which has a wide, but usually unrecognized, technological component, perhaps the topic of half-timbered houses could be considered. This topic occurs in almost all primary and middle/lower secondary schools within the scheme of work in History, either as part of a study of domestic architecture, or as part of a study of a particular period, but the work covered is usually limited to sketching an example of the style of architecture or perhaps of painting a cardboard box to represent the style. A few lucky pupils will visit a house of the period if there is one in the locality, but the benefit obtained from the visit will largely depend on the quality of the teacher's own preparation. In fact this topic is rich in technological applications if the house is seen not so much as a static building but as a machine for living in and as a technological development of its time and place.

Whilst not all of the aspects which follow would be included in the pupils' investigations, the teacher could encourage groups of children to follow different lines of enquiry, pooling the sum of their knowledge in a topic folder or a presentation of some

62

sort. Individual children showing particular enthusiasms could be allowed to follow these knowing that ultimately all will benefit from the sum total of the activity. Wherever possible the work should include practical activity because the value of making and doing cannot be overstressed, and this could include the collection, handling, investigation and classification of artefacts and materials as well as the making of models, in that all of these hands-on activities help to consolidate the acquisition of both knowledge and skill.

A useful line of enquiry, rich in technological possibilities, would be to take the aspect of *materials* and *constructions* used in the original period and to compare and contrast them with their modern counterparts. The sources of the materials used, and the methods of manufacture and construction, could be investigated and lists of tradesmen and craftsmen involved in the processes of housebuilding, then and now, could be drawn up. A model house could be made, preferably by making a balsa wood frame and infilling the spaces, perhaps with offcuts from polystyrene ceiling tiles, but this must be based on an original period house and not on a twentieth-century builder's interpretation!

The materials found in the half-timbered or timber-framed house would include oak for the frames, floors and internal woodwork, ash or willow for the wattles, daub consisting of clay, straw, hair and cow dung, iron nails, hand-made bricks, straw or reed for the thatch, cut blown glass for small windows, elm or ash for the shutters, wool for wall hangings and, of course, water, sand and plaster. In some houses of the period traces of wall paintings can be found and therefore pigments could be added to the list.

Other aspects of the period to be investigated could include frames, joints, chimneys, roof vents, ventilation, half doors, brick manufacture and thatching techniques, and all of these could be compared with their modern equivalents including extruded and moulded bricks, breeze blocks, thermal blocks, concrete and reinforcements, stainless steel, galvanized iron, graded plasters, insulation board, plaster-board, decorative wall-boards, melamine plastic sheeting, polyurethane, expanded polystyrene, aluminium, woodscrews, float and rolled glass, double glazing, glass-fibre matting and resins, quarry and glazed tiles, drainage pipes, plastic rainwater goods and so on. Specimens of most of these items are easy to acquire and manufacturers are normally willing to supply trade and advertising literature giving details of specifications and processes.

Still on the topic of timber-frame houses *economic factors* could form the basis of a second line of enquiry when sources of finance and materials, availability of roads, bridges and transport, and even fashion, could be investigated leading to work on incomes, manpower, building skills, resources, roads, canals and including modern aspects such as heavy goods vehicles, railways, air freight, materials science and even touching on international trade. The work could even lead to questions being asked as to why there is a revival in the timber-framed method of housebuilding in the 1980s.

A third line of enquiry could investigate *social implications* which would include personal and environmental factors comparing and contrasting past and present living conditions and the catalogue of topics could include domestic standards, fashions, tithes and taxes, size of family, status of occupier, availability of servants or domestic help, water supply from streams, ponds, springs, rivers and wells, drainage, disease, diet, clothing, furniture, rush lights and candles, peat and logs for fuel and also perhaps, transport. This could then lead on to topic work on modern public services, rates and taxes, hygiene, synthetic fibres and fabrics, electricity and gas supplies, other fuel sources including, perhaps, solar energy, dish-washers, refrigerators, deep freezers, washing machines, spin-driers and other domestic equipment and even video and home computer implications.

Obviously only a few of these topics would be included in the work with any one group but every one of them would imply an investigation of technological aspects relevant to the children's own lives and would cover subject implications relating to History, Geography, Science, English, Mathematics, Art and Craft, Environmental Studies and Health Education but giving a common purpose for the work.

An investigation of this sort can throw up interesting facts and lead to other lines of enquiry. For example, it appears that in order to make daub, the material which was plastered onto the wattles to infill the spaces between the parts of the timber frame of a half-timbered house, the builders heaped clay, straw, cow dung and horsehair in a field gateway adjacent to the house being built. A flock of sheep would then be driven backwards and forwards over the materials until all the ingredients had been mixed together. In effect it was a sheep-powered concrete mixer!

Perhaps some pupils in the class could be motivated by this quaint information to discover and investigate other machines or industrial aids, past or present, which depend on muscle power for their energy source. Another line of enquiry, as a result of seeing a hand-wrought nail, could be to investigate the Black Country nail-making industry. The information that water was drawn from the well in a leather bucket by pulling it up with a chain wound round a wooden roller could be extended to lead to experiments with paper chains, and chains made from paper-clips, with experiments on strength and an investigation of chain-making as a craft. Even the construction of chain-mail and its use in personal armour could come within the range of enquiry triggered off by the original activity.

Obviously, open-ended activities of the sort indicated in the previous paragraph need to be closely controlled by the teacher as it must be possible to bring all the strands together from time to time to keep control of the learning situation and to ensure that there is progression, that the children are being extended and that the work is being covered in sufficient depth. It must be realized too, that the work, if it is really child-centred, will appear under several subject headings.

Perhaps we should remind ourselves at this stage that 'subjects' as such are really academic conveniences for classification purposes and that they do not exist in real life. In order to achieve balance in the pupils' curriculum content the teacher cannot ignore subject content but there is no reason why the child should have to label an activity as being 'History' or 'Geography'. In fact, to do so can be counterproductive as it may affect learning concepts and reduce the benefits obtained from possible transfer of training situations. An illustration of this is that some years ago a young neighbour asked the author for help or advice with his homework. On showing the task he was told that he should have no problem in solving it as the example was a quadratic equation, something which he had frequently done at school. 'It can't be', replied the boy. 'This is Physics home-work, not Maths'!

Perhaps, at this stage, a few words about the process known as transfer of training would not come amiss because, although controversy still exists some seventy or more years after psychologists first raised the possibility of transfer as a direct result of the effects of training, it can be stated with some certainty that unless specific steps are taken to aid the process there is little likelihood of skills and knowledge acquired under one subject heading being applied or transferred to other areas no matter how closely they may be related.

In the teaching or learning situation it is essential that the child should be involved in a conative process, consciously striving for meaning and understanding. If the child's attention can be drawn also to the sensory stimulus of the activity including touch, smell, sound or appearance as appropriate, then the learning situation should be reinforced and there will almost certainly be greater understanding and better perception. Provided that the child understands the purpose of the activity, and that his attention is directed specifically to those aspects which have wider application, then there is an increased possibility that he will actively look for opportunities to apply this knowledge or to use the skills in a transferred situation.

A fairly simple example of this transfer of training can be demonstrated through the teaching of skills in tool usage. Perhaps a group of children is being taught for the first time how to use a junior hacksaw and the demonstration could start with the girls and boys being shown how to secure their work in the vice, how to hold the saw in two hands and how to obtain the correct stance. For this to be positive the children then need to get the feel of standing and holding the saw correctly and this would be practised under supervision with the teacher drawing attention to correct positioning and to the precise feel as, for example, the deployment of the hands with 'three fingers, one finger and thumb' in correct places and the way in which pointing the first finger of the hand towards the blade of the saw results in a stiffening of the wrist and therefore a more accurate cut. The children would then be shown how to use the saw to cut to a line and would be encouraged to discover how the saw blade actually cuts into the metal or plastic workpiece. Once this process has been assimilated and practised, transfer of training principles can be applied, perhaps in learning how to use a file. The pupils' attention is drawn to the fact that the stance and holding positions for the file are almost identical with those of the saw and that the cutting action also has many similarities. If the children are alerted effectively, not only will they acquire the skill of filing very quickly and efficiently, the practice in filing will actually help to consolidate skill in sawing and the two will be seen and understood as complementary skills with a large degree of overlap and there will be a generally improved facility and performance.

This principle of identifying and stressing general and specific factors, which appear to have relevance for other and alternative applications, can play an important part in the learning process, and can lead to more effective learning and higher standards of performance provided that the pupils are well motivated and are involved in the process with active understanding. If a linking theme can be found to underpin activities in the various subject areas, then the foundation for effective transfer of training can be laid. This approach will be developed further in the section on fields of interest.

The collection of activities included under the heading of Craft, Design and Technology has an advantage over other traditional and long established school subjects in that it covers no specific body of knowledge and therefore it can permeate all areas of the curriculum. Any activity which is concerned with the material and physical world or which investigates man's interaction with his environment comes under the CDT umbrella and, because it is such an all-embracing activity, it must play a major part in cross-curricular activity and can often provide the link or common ground between traditional subject activities and therefore may also create situations where transfer becomes possible (see Figure 4.1).

Earlier in this chapter it was stated that teachers need to exploit technological possibilities in their existing work in the classroom if there is to be a development of technological awareness on the part of the pupil. Fortunately, many existing curriculum activities can be seen to have technological relevance but it is also possible to introduce new facets to give the work a technological bias producing even greater cross-curricular potential but without making any drastic changes in classroom activities or practice.

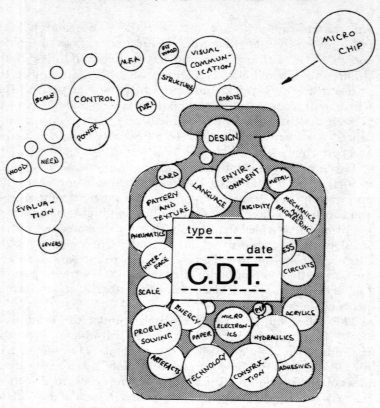

. . . . .covers no specific body of knowledge.

Figure 4.1 The ferment of CDT

It is proposed to look at some of the main subject areas in turn and to list activities which may be appropriate for a wide age range and which place the emphasis on technological awareness without any major deviation from existing approaches.

Geography: The technological connection in this subject area can come from work concerned with man's interaction with his immediate and wider environment, the exploitation and control of resources, and the ways and means whereby he has provided himself and his family with shelter and a means of support, and has steadily improved his standard of living. Comparisons between the industrialized countries and the ways in which the Third World countries are developing primitive technologies would also be most relevant.

Possible topics could include:

- *modifications to the landscape*:
  land reclamation, reservoirs, dams, irrigation schemes, paddy fields and terraces, tidal barriers, opencast mining, quarrying, tunnelling.
- *transport and communication*:
  roads, bridges, canals, railways, airfields, ports and docks.
- *private, commercial and industrial building*:
  housing, development of villages and towns, the 'new town' concept, shopping precincts, industrial estates, tower blocks, the garden village concept, the way in which local and national resources determine building styles and materials.
- *energy sources and resources*:
  fossil fuels, tidal and solar energy, power stations, atomic power, pressurized water reactors, power sources in the Third World countries, human and animal energy.
- *food production and farming*:
  mechanized agriculture, factory farming, fish farming, hydro-culture, the significance of the emergence of garden centres and the indoor garden.

History: This could be concerned with man's technological development through the ages and perhaps it could concentrate on his discoveries, inventions and achievements rather than on dates, battles and Royal families.

Possible topics could include:

- *tools, machines and equipment*:
  bone and antler tools, flint artefacts, bronze weapons and tools to the atomic warhead and nuclear submarine, smoke signals and the abacus to the space shuttle, the microcomputer and the radio telescope.
- *development of transport*:
  the pack horse and wooden wheels to Concorde and the hovercraft.

- *development of buildings*:
  house construction from the turf hut to the skyscraper; shuttered window spaces to the modern insulated home.
- *development of writing*:
  from the clay tablet through the quill to the ballpoint pen, and the typewriter to the word processor.
- *measurement of time*:
  from the sun dial through the hour-glass and water clock to the digital wrist-watch and atomic clock.
- *health and fitness*:
  medical advances and discoveries including anaesthetics, antibiotics, artificial limbs, bionic limbs, life-support machines, development of ambulance and firefighting services, sea rescue services, the Flying Doctor services.
- *aesthetic development of mankind*:
  primitive art forms and the influence of civilization and cultural development on art; exploitation of new materials in art form (for example, glass-reinforced plastics in sculpture and the use of resin-based paints to replace oil colours).

Mathematics: Whenever possible, practical applications of mathematical knowledge, concepts and skills should be undertaken relating mathematical ideas to other curriculum areas.
Possible topics could include:

- *measurement for a purpose*:
  dimensioned plans and maps, dimensioned sketches and working drawings, model-making, measurement in science and in craft activities, measurement in physical education and in music.
- *scale and proportion*:
  scale drawing and modelling, mapping craft activities.
- *spatial concepts*:
  sand and water play, painting and drawing, modelling.

Music: Music-making involves both scientific principles and craft and technological skills.
Possible topics could include:

- *making of simple untuned instruments*:

70

ideas about frequency and vibration leading to tuned instruments; principles affecting resonance; historical development from pan-pipes to the electronic organ.

Science: Science is concerned with man's exploration of the natural and physical world and the Universe in order to gain understanding and knowledge. This understanding and knowledge can then be used as tools in man's technological development going far beyond applied science which is normally concerned only with proving a theory.

Possible topics concerned particularly with craft, design and technology could be:

- investigation of materials and their properties
- physical laws
- gears and ratios, levers, screws, the inclined plane
- vacuum, hydraulics, pneumatics
- triangulation
- sound and light, the pinhole camera.

Art and craft: All work in craft, design and technology needs the support of work in art and craft. Aesthetic and manipulative skills should be continually developed in line with the physical and emotional development of the child and in the primary school it may be impossible to distinguish between the work in art and craft and in CDT.

Important topics which are relevant to CDT work could include:

- two- and three-dimensional representation of man-made artefacts
- model-making, especially of things that work
- problem-solving activity followed by evaluation
- joining of materials
- manipulation of materials
- constructional techniques and experimentation
- aesthetic and functional discrimination
- graphical communication as an alternative language.

Language development: One of mankind's greatest assets is its ability to use spoken and written language to communicate ideas, feelings and information and to bring order to human relationships. Words also enable man to engage in abstract thought and to pass information on from one generation to the next, thereby ensuring that development of the human race is progressive. Although there have been some developments in human relationships, social relationships have really changed very little over the centuries, although there is a veneer of greater sophistication, whereas technological understanding and ability has mushroomed, particularly during the lifetimes of the last three generations. This has only come about because of the continuous and cumulative effect of human knowledge coupled with the ability to concentrate on specialist interests and to work in groups resulting in achievements which total much more than the sum of the constituent parts.

The acquisition of language and the application of its rules is, of course, a steady and continuous process after the initial spurt which occurs in the pre-school period. Language development is in itself a cross-curricular interaction as much of the child's progress in this subject area comes about as a result of general curricular activity. If every opportunity is taken to increase language skills then progress will be with real purpose and will, in effect, be another example of transfer of training. Most new words, for example, are introduced through specialist subject activity and if this vocabulary extension is related to practical work then understanding and recall will be greatly improved. Technological activity is rich in vocabulary extension and if pupils are expected to research problems and then to work out solutions and evaluate the result there will be many opportunities for both descriptive and imaginative language skills to be practised.

Most new words that have to be added when a dictionary is updated have originated from scientific and technological activity and most of these words are used world-wide, based on their English forms. This means that teachers should endeavour to use correct technological terminology whenever possible with all age ranges and thereby reinforce and consolidate the children's technical and scientific language skills.

Possible activities that would ensure a technological bias could be:

- extending the vocabulary by using new words for materials, tools, processes, equipment and techniques;
- using books, pamphlets and journals as a resource for technological activity across the curriculum;

- practising factual reporting and the use of concise instructions in connection with technical processes carried out by the pupils, for example helping Mum or Dad:

>   mend a puncture
>   paper the kitchen
>   paint the front door
>   lay a carpet
>   dig the garden
>   build a rabbit hutch
>   light a bonfire
>   steer the shopping trolley
>   change a fuse
>   clean the foodmixer
>   ice a cake
>   cut the grass
>   hang a picture
>   emulsion the pantry
>   use the sewing machine

The writing of precise instructions might be related to such topics as:

>   drawing a cube
>   enamelling on copper
>   making a model dragster
>   making a kite
>   building a sand-castle
>   moving a heavy weight

The writing of descriptive prose or poetry will probably be largely confined to the specialist English lesson but topics could go beyond those of 'A day at the seaside with Grandma' type. For example, subjects could include:

>   making and firing a clay hedgehog
>   seeing a rainbow when heating copper
>   building the tallest tower with folded paper
>   flying a kite
>   taking a bicycle ride
>   the glow of the forge
>   the sound of planishing
>   rhythms on wood

how plastic is plastic?
does plastic have a memory?
Look! It works!

It may be true to say that technological awareness could not exist
without words as without them there can be little true understanding or insight and therefore language development at every age is
at the core of technological development. If only our present
generation of language teachers could have benefited from some
sort of technological bias in their own schooldays how much more
effective our present skills at language teaching might be!

Fields of interest

In the preceding pages on technology across the curriculum
reference was made to the use of 'fields of interest' to bridge
traditional subject boundaries and to provide a common purpose
to the pupil's activity, with Craft, Design and Technology
providing the core for wide-ranging subject involvement. This
appears to be an extremely valuable curriculum innovation but
what are 'fields of interest'?

In the 1940s and 1950s most junior schools introduced topic
work or project work into the curriculum and, in its most highly
organized form, it meant that a common topic was followed by
all pupils and classes in, say, a year group culminating in a
shared display in the school hall at the end of the agreed period
(normally a half or a full term). The intention was that the topic
would provide the motivation for individual or group work and
that although individual pupils would be working on different
aspects, each pupil would have opportunities to report back to
the remainder of the class resulting in shared knowledge and
experiences. In theory this would appear to have had much to
commend it but in practice there were difficulties and problems.
For example, the topic chosen for the term's work could have
been 'A Coal-mine'. In the sphere of Geography there would be
no shortage of subject matter for investigation, including distribution of mines and their types, varieties and types of coal and
their uses, road, rail and canal outlets and economic factors;
History would be concerned with the laying down of coal seams,
the development of the mine, the changing conditions for miners
and their families, improvements in mechanization and the

social life in a mining community; Art and Craft activity would include drawings, paintings, murals and model making; Mathematics could use problems connected with the purchase and delivery of coal by the kilogram and tonne but topics would perhaps be limited in this subject area; Scientific investigation could consider the composition of coal and the processes involved in making coke coal gas; English could use appropriate poems and prose about the mine with related written work and so on. Some subject areas, of course, would have no connection with the topic at all. This meant that as the subject involvement was not equal all sorts of distortions took place in order to stretch out the content of certain subjects to fit the overall time span. Rumour has it that it was not unknown, when the topic was about mines or mining, for the cookery class to be involved in the making of charcoal biscuits or burnt toast!

As the disadvantages largely outweighed the advantages the class project or topic gradually gave way to individual topics and these are common in the junior school stage up to the present day. With guidance and leadership, individual topics can be a fruitful source of learning provided also that adequate resources are to hand and that there is appropriate teacher intervention and control. All too often, however, there is little guidance in the topic chosen, much of the pupil's work consists of copying large sections directly from reference books and there is little to prevent the pupil from repeating the same topic year after year with different class teachers. An alternative system which has all of the advantages of a topic based approach and few of the disadvantages is to use a 'field of interest' as the motivation for both cross-curricular and specialist individual, group and class work. The field of interest should be broadly based so that it is not restrictive in terms of motivation and it should be used to exploit cross-curricular activity without expecting equal inputs from various subject areas. If the fields of interest are selected for their educational and technological potential then CDT once again forms the common ground linking the specialist subject interests.

The field of interest would normally be concerned with a concept or a property rather than with a specific object or situation and the choice would depend on the age and ability levels of the group or class. For instance, *refrigeration* would hardly be suitable as a topic for a group of 6-year-olds, but *floating* or *wheels* could provide both interest and motivation for that age group. The aim is to use the field of interest to provide the incentive and the common ground rather than to carry out an exhaustive investigation of the subject. If pupils' work under various subject headings appears to have a linking theme then not only will it be more purposeful but also it will provide greater opportunities for transfer to take place. Another difference between the earlier form of topic work and the present proposed approach is that there is no intention that it should be made to fit into a set period of time or that all subject areas should be equally involved. If connections are possible between the various areas of the subject work proposed and the field of interest then co-ordination should be encouraged but some subjects will not lend themselves to the particular theme and will continue to operate outside it. Sometimes it is mainly a question of timing in that proposed work can either be brought forward, or deferred, to exploit the possibilities which the process offers, but if co-ordination between subject activities and liaison between teachers occurs and if the pupils are motivated to carry out their work with enthusiasm and satisfaction then the process must be worthwhile.

In specialist teaching situations, a number of occasions arise where similar subject matter is covered separately under different subject headings because it has relevance for each of those subjects. This is wasteful in time and effort unless the duplication is deliberately included for purposes of consolidation and reinforcement and also it is highly likely that the child will regard the topic when covered by, say, Science, as being unconnected with the same topic when it is included in, say, Geography. Liaison between teachers could help to identify these areas of overlap resulting in more efficient use of time as well as more effective teaching.

Obviously, for the field of interest approach to be successful, considerable teacher preparation is essential as it is with any child-centred, open-ended approach. The teacher must research the topic to ensure that appropriate resources will be available and possible outcomes must be anticipated although this does not mean that the teacher must know all the answers in advance. There is nothing wrong in working alongside the children provided that potential targets are recognized and that support, encouragement and guidance can be given and that the children's horizons are continuously extended. Both specific and general aims and objectives must be clearly defined before any work is commenced and yet an open mind must be kept so that should the children's work open up new lines of enquiry then it must be possible to pursue these if it is deemed to be desirable. An example of this is shown in the DES film, *Practical Thinking*, where a Sixth Form pupil, in developing a kaleidoscope for use by handicapped children as part of an 'A' level Design course, becomes so engrossed in the shapes that he is creating that he ends up using those shapes to make a large-scale rocking horse toy for those same children's use and abandons his work on the kaleidoscope. Open-ended development of this sort is perfectly acceptable provided that it results from reasoned judgements, is progressive in nature, and does not allow the pupil to drift superficially and aimlessly from topic to topic to escape responsibility.

In the traditional classroom environment in the majority of junior schools the whole curriculum is normally in the hands of one teacher and therefore decisions about adopting the field of interest approach present few difficulties. In the open-plan environment found in most newer infant, junior and middle schools and in schools where specialized teaching occurs then liaison between teachers becomes very necessary. In an open-plan environment, because of the working conditions, there is normally a good deal of rapport between teachers working in the same base area and therefore liaison about curriculum matters is relatively easy provided that one of the teachers is prepared to take the initiative. In a school where teachers operate in enclosed boxes, and where the timetable is largely covered by specialists, then more effort and planning is needed if co-ordinated work with teacher liaison is to be introduced.

Any attempt at an imposition of this method of curriculum co-ordination is doomed to failure as participating teachers must be willing partners who recognize the possible benefits in motivation and avoidance of duplication of effort and who are prepared to exchange confidences. A better way to originate the change is for a couple of teachers to discuss the possible common ground over a cup of coffee in the staffroom then, when their own liaison pattern is working effectively, to begin to involve a third and then a fourth colleague. If there is a year co-ordinator then the process may be easier to introduce and progress will be quicker assuming that individual teachers are sympathetic to the proposals.

For most teachers the best way to operate field of interest activities will be by group work with either individual assignments in each group or with group assignments, depending on the age of the children and the complexity of the chosen topic. Group assignments are certainly easier to organize and control as a common source of resource material can supply the needs of the whole group. They also have the advantage that group activity can result in deeper understanding and a higher level of involvement and standards than is possible with children working in isolation. The teacher must be free to circulate around the class to guide and motivate children and to lead them on to new challenges. In some classrooms the 'crocodile' of children waiting for teacher's help is a common sight and one that should be prohibited! Not only does it waste hours of children's valuable time, but it also results in the teacher spending all of his or her time looking at work which has been completed, rather than moving around the class looking at work actually in progress and anticipating future needs and therefore, raising standards.

The use of short-term targets for each child, or group, will set the pace for child-centred activity and, above all else, the teacher must be alert to the potential of the work being undertaken in order to ensure progression and to stretch each child as far as abilities will allow.

The possibility of using *wheels* as a field of interest has been mentioned and, in fact, it is an area of investigation which has much to offer for all age ranges as it can trigger off many lines of enquiry and it is rich in practical applications:

Historical and geographical aspects:

- The development of the wheel through the ages.
- The wheel and its place in the development of civilization.
- The role of the wheel in the development of agriculture.
- The role of the wheel in the development of industry.
- The role of the wheel in the development of commerce.
- Application of the principle of the wheel to other machines, for example, treadmill, spinning-wheel, water-wheel, capstan, engine flywheel, gyroscope, clocks and watches.

- The mystery of the Aztecs and the Mayas who had wheeled toys but no wheeled vehicles.
- The wheel and tortures, for example, the rack.

Social aspects:

- Effect of the wheel and wheeled transport on all forms of social activity.
- Investigation of society's dependence on the wheel.
- What would life be like without the wheel?
- The wheel and recreation, for example, holidays, motor sport, cycling, fairgrounds, toys.
- The wheel and public transport.
- The wheel and public services, for example, water, electricity, gas, sewerage, drainage, refuse collection.
- The wheel in the kitchen, for example, food supplies, food distribution, food-mixers, food-processors, mincing-machines, hand whisks, oven-timers, cooker knobs.

Craft, Design and Technology:

- *Materials*: stone, wood, cast iron, wrought iron, steel, alloys, plastics, rubber, synthetic rubbers.
- *Construction*: solid, laminated, spoked, cast, moulded, pressed, machined.
- *Processes*: gluing, riveting, pressing, moulding, casting, turning, carving.
- *Parts*: hub, axle, shaft, spoke, rim, felloe, bearing, brake, shoe, disc, pad, tube, schrader valve, tread, fabric, steel wire, rim, tyre, solid, pneumatic, tubeless.
- *Twentieth century technology*: development of turbines, rotors, propellers, hovercraft.

The work could include investigation on axles, friction, ball bearings, braking systems, surface adhesion, surface pressure, alignment, bicycle spokes, tyre retention, methods of manufacture, wheels on railway trains and rolling stock, aeroplanes, armoured vehicles, farm tractors and farm implements, bulldozers and land levellers. Lots of fun could be a by-product as a result of making and testing models of all kinds, either made from a constructional kit such as Lego or Meccano or made from card or durable materials. Simple gear-wheels could be made by wrapping a strip of corrugated card around two or three circular

adhesives	air currents	balloons
boats & ships	boxes	bridges
buildings	bulbs and switches	castles
communication	compression	control
cylinders	electricity	electronics
energy	environment	expansion
falling	fire	flexibility
flight	floating	fossil fuels
friction	games	gearing
gravity	heat	inclines
industrial	inventors and	lamination
archaeology	inventions	landscape
levers	light	lubrication
mechanisms	movement	music
photography	power	printing
propellers	propulsion	radio
railways	refrigeration	roads
robots	rotation	screw threads
sinking	solar energy	sound
spanning a gap	structures	surfaces
tension	texture	transport
triangulation	tubes	tunnels
turbines	water-power	waves
wheels	wind	works

Figure 4.2 Possible fields of interest

boxes or drums of differing sizes (such as cheese portion boxes, goldfish food containers, paper drums and spice boxes) and these can be used to drive simple machines. Commercial wheels are available in Meccano, Fischer Technik, Lego Technical Functions and many of the plastic and wooden constructional toy sets and, also, wheels of various sorts can be obtained from model-makers' shops and suppliers.

With older pupils the propeller and the rotor such as are found on piston aircraft, helicopters, autogyros and hovercraft could be compared with the wheel and practical work could include the problem of changing linear motion to cyclical motion and vice versa (perhaps by making 'fun' machines).

The use of the wheel concept in astrological calendars, pie charts, clock faces and even traffic islands could be investigated as part of the study and, again, with older pupils, the wheelbarrow, the pulley and the block and tackle could be considered as part of physical science.

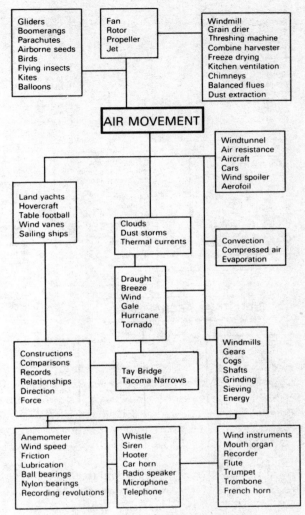

Figure 4.3 Fields of interest: air movement

Artwork could be used to consolidate the concepts gained, perhaps using a clock as the starting point.

Figure 4.2 presents possible fields of interest. Similar approaches can be used for most of the fields of interest listed earlier, as can be seen from Figures 4.3 to 4.6 which have been set out as flow diagrams for the sake of clarity.

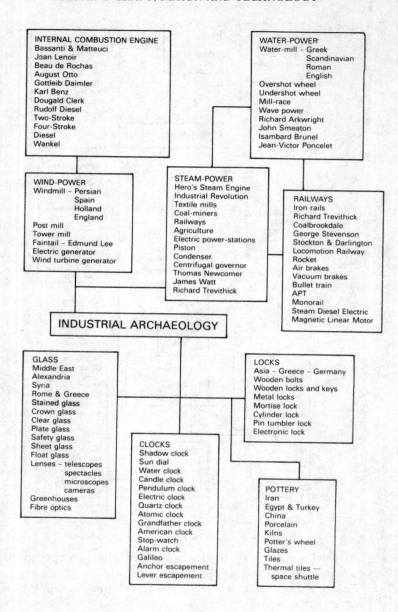

INTERNAL COMBUSTION ENGINE
Bassanti & Matteuci
Jean Lenoir
Beau de Rochas
August Otto
Gottleib Daimler
Karl Benz
Dougald Clerk
Rudolf Diesel
Two-Stroke
Four-Stroke
Diesel
Wankel

WATER-POWER
Water-mill – Greek
                Scandinavian
                Roman
                English
Overshot wheel
Undershot wheel
Mill-race
Wave power
Richard Arkwright
John Smeaton
Isambard Brunel
Jean-Victor Poncelet

WIND-POWER
Windmill – Persian
                Spain
                Holland
                England
Post mill
Tower mill
Faintail – Edmund Lee
Electric generator
Wind turbine generator

STEAM-POWER
Hero's Steam Engine
Industrial Revolution
Textile mills
Coal-miners
Railways
Agriculture
Electric power-stations
Piston
Condenser
Centrifugal governor
Thomas Newcomer
James Watt
Richard Trevithick

RAILWAYS
Iron rails
Richard Trevithick
Coalbrookdale
George Stevenson
Stockton & Darlington
Locomotion Railway
Rocket
Air brakes
Vacuum brakes
Bullet train
APT
Monorail
Steam Diesel Electric
Magnetic Linear Motor

INDUSTRIAL ARCHAEOLOGY

GLASS
Middle East
Alexandria
Syria
Rome & Greece
Stained glass
Crown glass
Clear glass
Plate glass
Safety glass
Sheet glass
Float glass
Lenses – telescopes
              spectacles
              microscopes
              cameras
Greenhouses
Fibre optics

LOCKS
Asia – Greece – Germany
Wooden bolts
Wooden locks and keys
Metal locks
Mortise lock
Cylinder lock
Pin tumbler lock
Electronic lock

CLOCKS
Shadow clock
Sun dial
Water clock
Candle clock
Pendulum clock
Electric clock
Quartz clock
Atomic clock
Grandfather clock
American clock
Stop-watch
Alarm clock
Galileo
Anchor escapement
Lever escapement

POTTERY
Iran
Egypt & Turkey
China
Porcelain
Kilns
Potter's wheel
Glazes
Tiles
Thermal tiles —
    space shuttle

Figure 4.4 Fields of interest: industrial archaeology

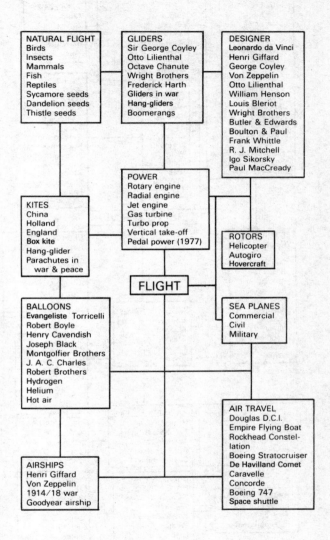

Figure 4.5 Fields of interest: flight

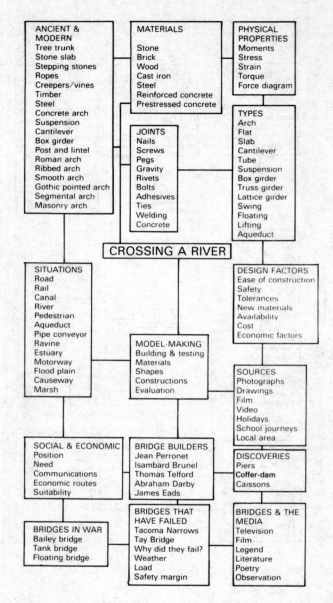

Figure 4.6 Fields of interest: crossing a river

Chapter Five

COMMUNICATION SKILLS

Children should be encouraged to record the outcomes of their observations as a first step towards developing a solution. This may involve drawing or writing, or the use of other media such as tape recording or photographs. Such records help children form hypotheses and promote imaginative and inventive solutions to their problems. Discussion of this material with others again plays an important role.

(National Curriculum Report: Science for ages 5 to 16: DES 1988)

GRAPHIC COMMUNICATION: AN ALTERNATIVE LANGUAGE

One of the most neglected areas of the curriculum in the junior and middle years phases is that of non-verbal communication. Very few teachers have had any training in graphics and they therefore lack both the confidence and the knowledge to encourage their pupils to use drawing as an alternative or supplementary means of communication in order to complement developing skills.

In the infant and first school, whilst children are at the pre-reading stage and are gaining control over the spoken and written word, drawing and painting play an important part in their development and teachers encourage self-expression through the use of crayon and paint. The children's drawing and painting consists of symbols which, apart from body movement and a limited speech vocabulary, are the only means of communication available to them. The sun, our house, mummy, daddy, trees, our car and other things connected with the child's close environment are represented by stylized symbols emphasizing those aspects which are important to the child. Trees are invariably drawn as a green blob on a brown stick; houses have a ridged roof, a chimney, four windows and a central door; the sun is always drawn as a yellow disc with rays emanating from it and the important parts of mummy and daddy are the eyes, mouth, hair and hands. Each picture actually contains a wealth of information and knowledge and it seems strange that some early years teachers appear to fail to recognize the benefits which could accrue from the encouragement of drawing and painting as an alternative language when they are presented with such clear evidence.

Even in skilled hands written and spoken language is a very imprecise means of communication, as is confirmed by the lengths to which legal experts have to go to ensure that their documents are precise and unambiguous, and a lengthy, wordy description is difficult to absorb and comprehend as it is essentially linear in form and has to be read and reread to associate later with earlier passages.

Language can become a screen which stands between the thinker and reality. This is the reason that true creativity often starts where language ends.

(Arthur Koestler)

A drawing, however, can be comprehended more quickly and three-dimensional representation adds realism and understanding to the communication. Annotation can be used to add to the content and, if necessary, phased drawings can show a sequence of events or stages enabling a process to be absorbed step by step.

Words are symbols too, but symbols only for sounds and those symbols have to be converted back into sounds, either mentally or vocally, before they can acquire meaning. Even then the symbols have to be taken in context if they are to have real and precise meaning. For example, the word 'elephant' conjures up a picture but of what sort of elephant? African? Indian? Adult? Infantile? Captive? Male? Female? Browsing? Charging? Many words have to be used if the image to be created by the reader is to resemble the image recorded by the writer. Even then there can be

86

misunderstandings if the reader's knowledge differs from that of the writer. In a similar vein, how on earth can an understanding of the concept of 'dogginess' be created without resource to the use of drawings, photographs, models, or a visit to the kennels? It is certain that an extraterrestrial visitor would have difficulty in establishing a concept of dogs as a species if given a list of different breeds or even written descriptions of some of them. He would find it equally hard if he was shown a Pekinese and a Great Dane. Given a number of drawings, however, he could deduce the common elements and begin to establish an appropriate concept. Drawings, in fact, would aid him more effectively than photographs as the very act of drawing necessitates extracting information and concentrating on basic points, whereas a photographic image contains much unnecessary information.

If words have been described as 'frozen sounds' perhaps drawings could be described as 'frames in a moving film', as individual sketches often capture a moment in a sequence of events. It would therefore seem that a drawing could lead to the viewer envisaging earlier and later 'frames' in addition to the one seen, a process which is very easy to create visually in a concise, compact manner.

In the same way that earlier periods are referred to as 'The Stone Age', 'The Bronze Age' and 'The Industrial Age', so the present and emerging period could be called 'The Communications Age'. The advent of modern printing methods, colour television, the computer, the word processor and other forms of electronic communication has resulted in massive attempts to capture the attention and interest of the consumer, and our children are just as much under pressure from industrial, commercial and political advertising as is adult society. Graphics have become big business, but not without reason, because there are few, if any, better ways of attracting attention and conveying information. The commercial artist of the 1940s and 1950s had a slogan which decreed that to be effective an advertising poster had to be so designed that 'he who runs may read'. In the 1980s the poster may be aimed at a passing motorist or at the family in the sitting room, but the principle remains the same in that the advertisement must be eye catching, interesting, and, above all, quickly and easily understood and remembered.

So it is with graphics in schools, as the aim is not to produce 'works of art' but to convey ideas or information simply, clearly and precisely. Some forms of drawing will be to convey ideas to other people, (the teacher perhaps), or to confirm that a principle or function has been clearly understood, but others will be to capture the pupils' own thought processes or ideas, and it is the latter activity which predominates in problem-solving activity. In investigating a problem and in searching for possible solutions, or variants on a solution, the pupil will produce ideas, sometimes fleeting in nature, which will be forever lost if they are not captured in sketch form. This means that the drawing, at whatever age or stage of development, must be simple and concise and, in the majority of cases, will consist of a few pencil lines conveying the idea of shape, function or the relationship of one part to another. Normally at this investigatory stage these different aspects would be shown on separate sketches rather than on one elaborate drawing as the aim is to capture and convey information rather than to present solutions.

If a local carpenter is asked to quote for fitting a shelf in an alcove in the dining room it is most likely that he would explain his proposals by drawing on the back of a piece of glasspaper or on an old envelope, and it is this sort of quick sketch that is under discussion in this chapter. Even allowing for his enormous artistic capabilities, Leonardo da Vinci's notebooks contain drawings which are not very different from those produced by children during problem-solving activity as he recorded only the barest necessary information in order to develop his ideas without embellishment.

In the design or problem-solving process it is mental images rather than words which enable objects, machines or mechanisms to be developed. Of course words play their part in helping to define the results and to provide the 'shorthand' in conveying the ideas to others. For example, a pupil's drawing of the blade of a wind-operated toy may have the word 'plastic' printed beside it as certain aspects cannot be established visually. Even so, many other words could need to be added (for example, 'rigid', 'flexible', 'coloured', 'opaque', 'thermoplastic') for precise meaning to be conveyed. However, the basic ideas and the relationship of component parts are normally created by mental imagery and, in fact it is possible, even normal, for mental manipulation of the images so that whole processes and sequences can be mentally

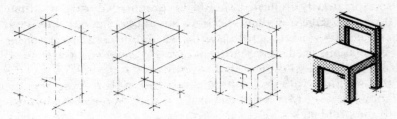

Figure 5.1 Outline drawing

'visualized'. For example, try to visualize the processes involved in pouring a cup of tea, or in extracting a match from a match box and then striking it. Sequential actions of this sort can be conjured up in a series of mental images and it is possible to take the process a stage further by imagining hypothetical sequences which could follow the known actions. It is therefore possible to test out theories by visualizing the possible outcome. Unfortunately these images are often transitory and very difficult to recall or to recreate and hence the need to capture key points with simple, quick drawings. In the process the act of drawing helps to clarify the idea and enables further imagery to be more precise and relevant.

For those readers lacking confidence in their own ability to draw, the first stage is to imagine the packing case, or packaging, which would contain the object to be drawn. A few pencil lines will produce the outline of this packing case and then it is relatively easy to draw the object within these guidelines. Having created the outline of the object details can be filled in to create a 'three-dimensional' drawing. No embellishment or unnecessary detail should be added as the secret of success is to convey key information only (see Figure 5.1).

A variety of sketching styles can be developed by using different media (ball point, fibre tip or felt pens, water-colour markers and wax crayons all give different results) and it is worthwhile experimenting to discover which drawing instrument suits your own style. Pupils will often produce interesting results by using more than one type of drawing instrument in a sketch and this technique can be used to highlight important details. In problem-solving activity lots of quick drawings of the sort previously indicated should be produced during the first divergent stage and then selections should be made and refined and details worked out during the second or convergent stage. There is always the temptation to latch on to a preconceived idea and to try to avoid the initial investigatory stage but, of course, this has little value as without investigation there is no way of knowing if the idea is suitable or if it is the best possible solution to the problem set.

Some items are rather difficult to draw, particularly if they are shaped in more than one plane (for example, a set of laminated salad servers), and it is sometimes better to use scissors and paper or card rather than pencil and paper. Similarly, paper and card mock-ups can at times be used instead of, or in addition to, free-hand sketches.

In cross-curricular activity, or in general classroom subjects, drawings can be used differently as the aim here would be to aid communication and to supplement written work rather than to solve problems. Imaginative drawing would be appropriate in creative writing and, in fact, would help to increase imagery and therefore lead to an increase in vocabulary. In the same way that mental images can be created and captured in problem-solving activity when dealing with functional and mechanical objects so can images be 'frozen' when the topic is concerned with events or attitudes rather than objects.

Perhaps more time should be spent in the classroom investigating the sort of instructional leaflet which accompanies most consumer durables when they are purchased and then producing similar guidance brochures either for hypothetical products or for actual items (toys, perhaps) which the children own or have made. The strip-cartoon style of layout is particularly effective for this type of communications brochure as each frame leads directly to the next and a sequence of events or stages can be illustrated. It is a good idea too to take a set of written instructions and then attempt to portray them visually and, vice versa, attempts can be made to translate illustrated directions into written form and to draw conclusions about the relative effectiveness of the different methods of presentation after evaluation has taken place.

Graphs and pictograms are other forms of visual communication and if to these are added charts and flow diagrams then it can

A B C D E F G H I J K L M N O P Q R S T U V W X Y Z 1 2 3 4 5 6 7 8 9 0

Figure 5.2 Simple line alphabet

be seen that visual or graphic communication can take many forms but all of them are concerned with giving information, or transmitting ideas, simply, clearly and without ambiguity.

It has been mentioned earlier that in many cases a combination of drawings and captions will give the clearest information and it is important that the lettering used is as simple, clear and unambiguous as the drawings of which it forms a part. The style used is not too important provided that it complies with the guidelines above but it is strongly suggested that the clearest upper case (capital) letters are those based on the proportions of the Roman alphabets. Perhaps copies of a simple line alphabet could be duplicated and displayed in all classrooms or teaching areas and middle years children could be encouraged to use this style of lettering whenever they need to use upper case letters in their day-to-day work (see Figure 5.2). In terms of layout, of course, any lettering should be seen as an integral part of the drawing and should be strategically placed so as to refer to the appropriate detail or area.

Sometimes the lettering will take the form of explanatory notes in which case clear handwriting or typed notes could be used and set in panels either at the side of, or immediately below, the relevant drawings.

In an environment dominated by the media where the child is surrounded by signs and symbols at home, at school and in the street, it is vitally important that all boys and girls should become proficient at graphic communication and should regard the activity as a normal part of the information seeking and transmitting process.

Language skills in Craft, Design and Technology

In the section 'Technology across the curriculum' in chapter four, the possibility was discussed of introducing a technological bias into the teaching of English and the development of language skills. In the same vein it has always been accepted that 'every teacher is a teacher of English' and this certainly applies in the teaching of CDT. The subject is rich in processes, tools, materials, activities and observations providing motivation for self-expression and the need to communicate, and every opportunity should be taken by the teacher to exploit the ongoing situation.

There are frequent references in this book to the importance of visual literacy but communication through graphical symbols is a complementary or alternative activity and not a substitute for the spoken or written word. In fact, it may be that we see only those objects for which we have a name, and our understanding of the environment may be limited to that which we can describe by name or by comparison. That is, we can sometimes overcome the problem of not knowing an object's name by comparing it with something known and familiar but it could be that objects and conditions for which we have no names, for us, do not exist.

An illustration of this could be that in Great Britain snow is an occasional nuisance but flowers are with us in abundance all year round. In the Arctic there is snow in abundance and flowers are few and far between and of no importance to the native population. The result is that an Englishman has few words for snow (classifications are limited to snow, slush, sleet and hail) but hundreds of names for flowers, whereas the Eskimo has dozens of words describing the state of the snow mantle but one name only to describe any class or description of flower. Could it be that scientists are prevented from recognizing the potential of some of their observations because there are no words to identify the phenomena? Certainly the growth which became known as penicillin had been observed countless times before its possible medicinal value was recognized. Is the American Space Programme hindered in its exploration of other worlds by human inability to recognize anything which has no named counterpart on our own planet?

Be this as it may, it is certain that words aid both definition and recall, and the wider the vocabulary the more precise the quality of the communication and the greater the understanding. For instance, compare the following:

'Please pass me the hammer'.
'Please pass me the Warrington hammer'.
'Please pass me the Warrington no. 2 hammer'.
'Please pass me the Warrington no. 2 hammer with the Hickory handle'.

Obviously in a conversation there is no advantage in the speaker knowing all the right words if they are not also familiar to the listener and there is a lot to be said for the infant school technique of labelling items in common use. In the early years classrooms large labels proclaim 'Door', 'Window', 'Table' and a similar process could be invaluable when teaching CDT. The toolrack in the cupboard, trolley or drawer could be labelled as should also each type of tool. The Dymo tape system is good for this or other types of self-adhesive labels could be lettered and applied. Some tools could have their names stuck to their handles and/or to their storage places and all tools should always be referred to by their correct and full names. Similarly, specimens of the materials to be used in the classroom or workshop should be labelled and be either on display or available for reference purposes. During the investigatory stage of any problem-solving activity these specimens should be regarded as an essential resource.

Again, in the infant school, most classrooms have a display corner or table where items with a common feature or property are displayed for a few days at a time. The theme may be 'silver' or 'shiny' or 'soft' and individual items are often labelled. What a pity that a display similar to this is not seen more often in the junior, middle and secondary phases as it has much to commend it as an aid to observation, understanding and literacy. The display may be of British hardwoods, where examples of oak, ash, beech, elm, cherry, lime and sycamore, all labelled, may be shown, or it could be of 'gears' or 'gearing' displaying items such as a clock, food whisk, hand drill, bicycle and pencil-sharpener. At times a single item could be put on display with its associated words connected to appropriate features with thread. For example a small table could introduce words such as the following:

Games table, telephone table, nest of tables, occasional table, coffee table, Long John table, veneer, plywood, Japanese oak, quarter-sawn, slab-sawn, grain, medullary rays, figure, pattern, heartwood, seasoned, kiln-dried, stable, shoulder line, tapered, mortice, tenon, adhesive, rail, stretcher-rail, bevel, chamfer, inlay, banding, surface, polish, varnish, lacquer, stain, gloss, satin, sheen, resistant, machine-made, buttons, fixing plates, brass counter-sunk woodscrews, slots, flush fitting, and so on.

and the more of these features that can be recognized, the greater will be the observer's understanding, not only of that particular table but of tables in general. Obviously, for this to happen there must be a definite and firm connection between the feature and the word and there is need for consolidation and reinforcement by teacher intervention and follow up. If an area of the room is to be reserved for a time for a certain activity, then the key words associated with that activity could either be displayed in the area or be listed on a workcard issued to the pupils. At times, of course, these words can be researched by the pupils themselves, but the teacher must ensure that the needed information is available and help must be given with lines of enquiry.

When engaged in practical activity pupils must be encouraged to describe not only what they see but also what they feel. Sometimes this will involve written work, but more often it could be a verbal description from one child to another, especially if they are working in partnership and sharing the experience. What does it feel like to saw through a piece of wood, to hit a nail with a hammer or to feel the rough side of a sheet of glasspaper? What are the child's personal feelings when it is discovered that the model windmill just completed really works or that the model dragster will not race forward because its wheels spin on the polished floor?

Similarly, as part of the problem-solving process, pupils must be encouraged to describe accurately what it is that requires attention. When are the dragster's wheels slipping? What are the possible or probable causes? Which possibilities are most feasible or easiest to try? What alternative materials are available? Are there any physical laws which explain what is happening?

Recent educational reports by HM Inspectorate and others

have indicated that perhaps too much time is spent by pupils in unproductive writing and copying and certainly many teachers feel that a lesson is incomplete unless the pupils have produced some written work. Perhaps one of the reasons why many British pupils are relatively inarticulate when compared with their American or European counterparts is that we do not encourage informal discussion sufficiently in our schools in that we feel insecure unless we have a quiet and well-ordered environment. In CDT activity there is certainly the need for things to be well ordered – in fact it is unlikely that any real learning would take place without first rate organization and self-discipline, but an ideal opportunity exists for the encouragement of purposeful discussion between partners or individuals in a group about the work in hand, and properly motivated problem-solving activity is an ideal vehicle for shared learning experience and for the acquisition and practice of language skills.

The writing and drawing involved in preparing the design brief in problem-solving activity is an excellent exercise in communication skills as the pupil has to inform others whilst confirming his own ideas. The evaluation stage, too, makes demands on developing language skills. It may be a good idea to encourage the keeping of a progress diary during the 'making' stage and this could be incorporated in the final brief to record the trials and tribulations, disappointments and successes of the problem-solving process. Another important aspect of language skills in CDT is in the application of correct scientific terminology to design activity and it is vital that every opportunity is taken to relate scientific facts and knowledge to practical applications in this way whether it be sand play in the reception class or work on inclined planes with upper juniors. Technology relies heavily on physical science as a major resource, but it is important that principles and terminology are applied correctly as this not only improves language skills, but also assists the process of transfer of skills.

The majority of new words introduced into the language in the twentieth century have been technological in origin and many of them have retained their English form throughout the world. The space programme brought a wealth of new words and processes and now the world of computers and computer education is introducing many more and, or course, unless a school is fortunate enough to have really up-to-date dictionaries, many of the new words and technical terms will not be included. With this in mind it is well worthwhile considering the possibility of requiring each pupil to keep a CDT vocabulary notebook in which new words, phrases and expressions can be recorded. This will serve a dual purpose in that not only will it give the correct spelling of the new words but also the act of writing out the word when entering it into the notebook will help to reinforce learning. A large-scale chart of the week's new words would also help to reinforce and consolidate the learning processes.

The National Curriculum Report 'Science for ages 5 to 16' states that, for the 7 to 11 year group, their work under the 'Communicating Technology' attainment target should include 'experience [in] the use of a computer for word processing, and data handling'. The statements of attainment section also refers to the fact that this age group should be able to use information technology (IT) techniques to help when designing.

The Final Report of the National Curriculum Design and Technology Working Group (May 1989) stresses the importance of IT in enhancing learning at all levels across the curriculum. It involves identifying and developing IT capability, the co-ordination of pupils' experiences in order to draw out the knowledge, skills, understanding and values making up IT capability, and the construction of a framework for assessment, identifying progression in terms of width and complexity. It also states that 'during the first two Key stages pupils should develop an awareness of the variety of applications of IT and the ability to use some of these applications in their own work'.

The microcomputer, with its associated word processing, visual display unit (VDU) graphics and printing facilities, has opened up a whole new world in the potential for the communication of ideas and the rapid transfer of information and, right from the earliest days in school, pupils must be introduced to its potential.

Chapter Six

THE WORKING ENVIRONMENT

Within any particular form of curriculum organisation there are various ways of arranging the work: children may be grouped, taught as a whole class, or work individually. There are advantages in variety, particularly when children have the opportunity to investigate and to work co-operatively and to communicate their ideas and findings in a variety of ways.

(National Curriculum Report: Science for ages 5 to 16: DES 1988)

ORGANIZING THE CLASSROOM FOR CDT ACTIVITY

If CDT is to permeate the whole curriculum in the 5 to 13 age range, then it follows that most of the activity will take place in the classroom. Indeed, apart from the 9 to 13 middle schools, most establishments will have neither specialist rooms nor specialist teachers so there will be no alternative. Using a normal classroom for practical activity obviously presents problems if it is intended that the subject is to be timetabled with every child involved at the same time, problems not only in organization but also in supervision. In the specialist workshop pupil numbers are usually limited to twenty for educational as well as for safety reasons and, in practice, the number is often less. In the classroom, however, not only are there larger numbers but also the teacher is having to cope with makeshift provision whilst lacking specialist knowledge and skill in CDT. If cutting tools are being used the very fact that there is a relatively crowded environment will increase the danger and, whilst it is highly unlikely that chisels or other sharp-edged tools will be in use, a pair of scissors carelessly handled can be just as dangerous.

For every reason, then, it is preferable that practical CDT activity should be so organized that a relatively small number of pupils would be needing to use tools and equipment at any one time whilst the remainder of the class is doing other curricular work. Ideally, some sort of CDT activity would be in progress by various small groups or individuals for the greater part of each day.

If the CDT work is related to the work of the remainder then, of course, the curricular connections will be obvious and the work will be enhanced. Much of the classroom work in CDT will be in the realm of technological awareness, and it is probable that much of this work will be non-practical in nature but, of course, there will still be a need for adequate organizational provision if any degree of child-centred learning is to take place.

The final nature of the provision will depend partly on the age and, therefore, size of the children, and partly on the abilities and interests of the teacher and the range of the curricular experience available in the school. However, certain things can be seen as basic requirements and it may be desirable to discuss these first.

For any sort of meaningful technological awareness to be developed it is essential that there should be adequate literary and pictorial resources available. This could be in the form of magazines, books, brochures, cuttings, manufacturers' and suppliers' literature, teachers' and commercial workcards and worksheets, material on loan from the library service and selected examples of the present or former pupils' work. It takes time and energy to build up an effective resource bank of this kind and it is a task which never ends as there is a constant need for updating, reclassifying and making additions to the range and, quite honestly, it is a job which can be done much more effectively by a group of teachers working as a team than by one person working in isolation (and often duplicating the work of others).

For a resource system to work effectively with the minimum of teacher intervention a simple and easily understood system of classification and retrieval is essential. This must be linked with an adequate storage system which will enable alterations and additions to be made as material becomes available and as needs change. Box files, shirt boxes, loose-leaf files, manilla or other strong envelopes, home-made sugar paper folders or pockets, even cornflake packets all present possibilities, and the final choice will probably depend on need, storage space and financial constraints. Chapter seven on Resources for Learning develops these ideas but at this stage it cannot be stressed too strongly that adequate resources are essential if cross-curricular technological activity is to take place.

In order to cope with practical and three-dimensional work, especially if it involves manipulation of the more durable materials, some modification is needed to the normal provision of desks or light tables. If finances and space permit, a general-purpose craft bench of the type designed for middle years use is ideal as it provides vices for work in wood, metal and plastics for four pupils, as well as offering a flat-topped surface when needed for general craft work or for graphics activity. Other, smaller, two-place benches are available from leading art and craft suppliers and from specialist bench manufacturers and these may be more suitable for use in infant and first schools. An alternative solution is to acquire a fairly heavy table of the type used in Home Economics rooms and these are sometimes available from the LEA's store of surplus furniture. A vice of some sort is essential so that pieces of wood, metal or plastic can be held securely, leaving both hands free, and if a table is being used a lightweight clamp-on vice is available which is more than adequate. The bench must be sited where it can be supervised and the practice, especially in infants' schools, of allowing children to work outside the classroom 'because I cannot stand the noise of hammering' should be discontinued. Apart from the need to be able to see the children at work for safety reasons, it is also very necessary to oversee the use of tools at all times to ensure that correct methods of handling and application are being used. No child should ever be left to its own devices to find out how to hold a tool or how to make it work. These things must be taught, and taught correctly, and there must be follow-up supervision whilst the skill is being developed.

For work using less durable materials ordinary desks or tables will suffice, but it is desirable to cover the tops with newspapers or a sheet of plastic to protect the surface from paint and scratches. If newspapers are being used it is advisable to place them on the tables upside down and to remove any copies of page three (unless language development is of prime importance!). An even better cover could be made from a sheet of hardboard cut to size to fit over two or four tables making a large, flat, working surface. This could be further improved by making it initially slightly oversize and adding a softwood lip (approximately 15mm × 15mm) by screwing it under the outside edge to locate it on the tables and also to protect the edges of the hardboard giving it a longer life.

It is a good idea to reserve one set of desks for practical work if a specialist bench is not available, as then a wider space can be left around this area making movement easier and reducing any problems which could be caused by congested conditions. Naturally, a specialist bench would be similarly sited if space permitted.

If the classroom has a side bench or cupboards with a flat working surface then this may be suitable for practical work if it is strong enough to withstand fairly heavy use and it would have the advantage of being slightly away from the rest of the desks or table creating a less crowded environment.

Remember, too, that practical work creates the need for storage and display space. Work in progress needs protection to avoid accidental damage and even an empty cardboard box placed over the models can make the difference between success and disaster if storage space is at a premium. Damage to the childrens' work should be at an absolute minimum as pupils should be encouraged to show respect for one another's models as part of social education, and if the CDT work is related to the other curricular activities going on in the classroom then there should be a group spirit and greater mutual respect by the pupils for all the work in progress.

The tools needed in a classroom situation would depend on the age and physical size of the children and on the skills and interests of the teacher, but they should be adequate to enable a wide range of durable materials to be explored and manipulated.

For infant and first schools catering for children between the ages of 5 and 8, a suitable list of tools could include:

bench hook
brush and dustpan
bradawl no. 1 or 2
gent's saw 100mm
glasspaper block

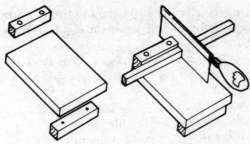

Figure 6.1 Bench hook

junior hacksaw and blades
mallet 100mm head
pincers 100mm
pin hammer
pliers: engineer's
pliers: round nosed
screwdriver: cabinet 150mm
screwdriver: electrician's 100mm
surform block plane
surform shaper
tack hammer
wheelbrace
   3mm, 5mm, 6mm HSS drills
hole saw (disc-cutter) for teacher's use

The bench hook could either be purchased or made from offcuts of timber. It is used to hold the wood safely and firmly while it is being sawn and it also prevents damage to the bench or table from careless use of the saw. Ideally it would be made from beech wood although other scrap wood could be used. Two strips each about 150mm long and 25mm square plus one flat piece about 200mm long, 175mm wide and 25mm thick would be needed together with four steel countersunk-head woodscrews 37.5mm long by size 6 or 8.

One of the 150mm strips should be screwed across the flat surface of one end of the larger piece and the other strip screwed underneath at the other end creating a sort of hook as in Figure 6.1.

In use, the bench hook is placed up against the edge of the bench and the workpiece is gripped firmly against the top strip with three fingers over the strip, the forefinger on the top of the strip and the thumb pressing against the side of the workpiece so that it cannot move.

In junior schools and in middle schools catering for 8 to 12-year-olds a list of tools for classroom use could include:

Abrafile (handled)
bench hook
bradawls no. 1 or 2
brush and dustpan
centre punch
coping saw and blades
files: half round 150mm and 200mm
files: second cut 150mm and 200mm
files: smooth 150mm and 200mm
glasspaper block (Cork rubber)
hammer: ball pein (peen or pane) 250g
hammer: pin
hammer: tack, model-maker's
hammer: Warrington 250g
hole saw (disc-cutter)
junior hacksaw and blades
mallet 100mm head
marking gauge
marking knife
nail punch
plain brace 200mm
    improved centre bits 6mm, 9mm, 12mm, 18mm, 25mm
    rosebit: square shank
pincers 125mm
pliers: engineer's 125mm
pliers: long nosed 100mm
pliers: round nosed 100mm
saw: dovetail 200mm
saw: gent's 100mm
screwdriver: cabinet 150mm
screwdriver: electrician's 100mm
screwdriver: electrician's 150mm
screwdriver: cabinet 150mm
steel rule 300mm

Surform block plane
Surform planer file
Surform round file
Surform shaper tool
try square 150mm
wheelbrace
   HSS drills 3mm, 5mm, 6mm
   countersink bit with round shank

Even in 9 to 13 middle schools where there is a specialist teacher and a CDT workshop there is still need for CDT activity to take place in the classroom areas to supplement the specialist work, particularly during the first two years, and to provide the opportunity to use the more durable materials in cross-curricular activity. It is therefore desirable to provide an additional set of tools as per the 8 to 12-year-old's list which can be used in any of the year bases under the direct supervision of the class teacher. If the tools are fitted into portable trays or mounted on a small trolley they can be moved from place to place as needed. In this situation the teaching staff would have an advantage over their primary school colleagues in that the specialist CDT teacher could provide in-service help and set standards and methods for skills and processes.

Any tool provided must be stored properly and looked after with respect if it is to give long and efficient service, and it is highly desirable that some sort of storage system should be contrived so that each tool has a home where it cannot rub against others. There is nothing worse than to see a collection of expensive tools thrown into a drawer or box and not cared for properly.

Wooden blocks or spring tool clips can be used to secure each tool safely in a drawer or tray or on a sheet of blockboard and if the silhouette of each tool is cut from thin coloured card and glued underneath the appropriate tool then every place will be easily recognized and it will be obvious if any particular tool is missing. Using Dymo tape or a similar lettering device each tool space could be labelled, aiding recognition and adding to the children's vocabulary in the process.

In addition to a range of tools it is obviously necessary to have supplies of sundries. These could include:

assorted glasspaper
assorted round wire nails, especially 25mm and 37.5mm
assorted oval nails, espeically 25mm and 37.5mm
assorted countersunk head wood screws
assorted round head wood screws
assorted nuts, bolts and washers, especially
    2BA, 4BA and 6BA
water-based contact adhesive
PVA adhesive
rubber bands
thin string
a length of candle (for work on friction where washers are
    needed)
clothes pegs, paper clips and spring clips
    (useful for holding small items while gluing)
75mm G-cramp could be a very useful addition

It is very difficult to draw up a comprehensive list of appropriate
materials for CDT work as every type of material from paper to
concrete could be included provided that it is non-toxic but the
following could be useful:

knot-free softwood (Parana pine or similar)
easily worked hardwoods (lime, jelutong, obeche or agba)
balsa wood (expensive but very useful where thin sections are
    required)
expanded polystyrene (if it is to be cut with a hot wire then
    ensure that it is non-toxic)
acrylic resin offcuts (Perspex, etc.)
soft-iron galvanized wire
plywood offcuts
hardboard offcuts
insulation board offcuts
cardboard of various thicknesses
cardboard boxes and cartons
plastic containers
'squeezy' bottles
thin garden cane
dowel rod
cotton reels
paper straws

cardboard tubes
corrugated card
ice lolly sticks

In all types of schools construction kits appropriate to the age
and manipulative ability levels of the children should be available. In schools with very young children the large plastic and
wooden kits with large plastic nuts and bolts or with click-
together assembly methods are ideal. Large wooden blocks and
even odd offcuts of wood are useful for random building and
construction experiences.

With rather older children constructional kits based on sound
engineering principles should be provided and these could
include:

Lego Technical Functions
Fischer Technik
Meccano
Stokys
Electronic workshops (battery-operated)
Electrical circuitry kits (battery-operated)

Finally, from time to time the classroom should contain bicycles
or bicycle parts, domestic equipment such as hand whisks, tin-
openers of various types, nutcrackers and anything else which
could excite the pupils' natural curiosity and help them to apply
theoretical principles to the everyday things of their environment. Having discovered the relevance of levers, gears, pulleys,
and so on, they should, of course, apply these principles to their
own models, making things that really work.

WORKSHOP ORGANIZATION FOR THE MIDDLE YEARS

Some middle schools catering for the 8 to 12 years age range, and
almost all 9 to 13 middle schools, are designed to include
specialist provision for Art and Design, Science, Home Economics, Fabric Crafts and Craft, Design and Technology. Hope-
fully, the architect will also have designed the school so that there
is a natural flow between these subjects, as without a distinct
physical relationship of one area to another it would be difficult
to sustain the necessary interface between the different subject
activities (see Figure 6.2).

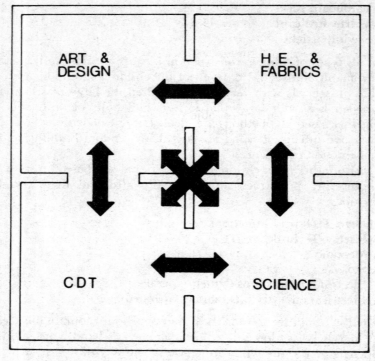

Figure 6.2 Liaison between specialist subject areas

In order to encourage cross-curricular activity and special relationships, and therefore more meaningful activity, within this overall area there needs to be visual access between the subject areas so that they are seen to be interrelated, and this can be achieved either by providing fairly wide archways instead of doors, or by designing the areas so that there are glazed openings between them. Careful siting of the archways can provide areas which 'contain' and baffle any noise so that one type of activity does not impinge unduly on any other. Of course this sort of area can only be really effective if the specialist teachers co-operate and work in harmony. It would still be possible to create imaginary walls and barriers and to treat the areas as though they were built as traditional box-like classrooms but this would be a sad misuse of a very desirable provision.

In some schools common aims and objectives are identified and team-teaching methods are used with spectacular results. For the teacher who has come from working in a traditional classroom-based building this may be too big a step to take and it may take several years to develop the special attitudes and abilities needed for full team-teaching methods to be adopted. Perhaps the most which many of us could achieve would be to identify common aims and objectives but to relate these to the exploitation of our own skills and interests so that we work as members of a team (which is not the same as team teaching!). The CDT area must be equipped sufficiently comprehensively to enable the pupils to experience all those areas of activity which they would have covered by the age of 13 had they transferred to a comprehensive school two years earlier and, in fact, if the middle school uses its facilities and strengths properly they will have had four years of specialist CDT experience by the age of 13+, whereas pupils who transfer at 11+ will have had only two years of CDT experience and influence.

The middle school CDT teacher is fortunate in that he/she has no constraints imposed by the examination syllabus and therefore there is much greater freedom to respond to environmental stimuli, but this is counterbalanced by the lack of strength and physical maturity of the middle years pupils and therefore work must be carefully selected to maintain high standards of skill and craftsmanship and tools and equipment must be carefully chosen with the aims, objectives and realities of this age group very much in mind (see Appendix A for list of tools and equipment for the specialist CDT area). In the 9 to 13 middle schools the pupils should have the opportunity to participate in a wide range of activities and explore and become competent in the use of most of the common materials, and this means that comprehensive specialist facilities are needed in the CDT area. The room needs to be some 70 square metres in size, with a bay for heat treatment processes, and a generous enclosed, or partly enclosed, area for storage. Good natural lighting supplemented by fluorescent strip lighting is essential, and fume extraction will be necessary in the heat-treatment area which should be sited away from natural light so that the flame and colours on the metal can be seen. At least one side of the room should be fitted with side benching with cupboards underneath and the top should be of solid beech enabling engineering vices to be fitted when needed and giving a good, solid working surface. This side benching also needs access to both 13 amp and low-voltage socket outlets for work in technology.

The heat treatment area should be organized so that there is a natural 'flow' between processes and therefore as little movement by the pupils as possible. Soft and silversoldering processes should be adjacent to a Belfast type sink and the acid bath should be housed in a properly constructed lead-lined acid cupboard underneath the draining board. The floor in this area should be able to sustain hard knocks and be fire resistant. In the remainder of the workshop the floor should have a warm, non-slip surface which will resist burnishing from woodshavings and which will not be too unkind to dropped tools.

Appropriate machines are an important part of the middle years CDT workshop provision, but they will tend to be used mainly by the older pupils, for safety reasons, and therefore they should not be allowed to tie up too much of the available floor space. Some machines will be mounted on the side benching and the others should be sited around the room and parallel to the walls. At least a metre should be left between each machine and, when they are in use, there must also be a metre of clear space in which the pupil can operate. The centre of the floor should be clear of all attachments and equipment so that four or five middle years' benches can be installed, each catering for four pupils. The benches should be fitted with woodworking vices and there should be provision for adding small engineer's vices when metal or plastics are being worked. An additional facility would be to have flat covers which can be placed over the benches when design work or activities needing a large, clean area of working surface are in progress.

The CDT area should have at least three emergency stop buttons installed at strategic points so that all machines can be stopped quickly in the event of an emergency. It is a good idea to have a large red arrow pointing to each stop button to increase the visual impact and, of course, every pupil should frequently be reminded of its purpose. The teacher should check the buttons daily for correct operation and any fault must be reported immediately.

The electrical supply to the work area should be controlled by a key so that unauthorized use of machines can be prevented. Obviously, the key should be removed when the specialist teacher is not present.

Electrical tools which are held in the hand when in use, such as soldering irons, pokerwork and engraving tools and power tools generally, should be low voltage models and manufactured with double insulation protection. In a new school the architect can include the low voltage supply in his initial plans, but in older buildings a transformer will be needed to reduce the voltage to either 24 or 110 volts depending on the type of appliances recommended (see Figure 6.3).

Storage of materials requires considerable thought if items are to be to hand and easily accessible. For example, wooden boards store best in a horizontal stack provided that they are supported on a level surface but when stored in this way the whole stack has to be moved to get to a particular board. If stored vertically they are easy to retrieve but there is always the danger of a board falling over and causing injury or damage. The method recommended by the author is that the boards should be stored vertically in wooden or metal racks but that there should be drop catches across the front of the racks preventing displacement of the timber but giving easy access. These catches could be made from steel strip or bar as in Figure 6.4. Metal, rod, bar, tube and angles could be stored in a similar vertical rack although, of course, the rack could be much smaller because of the smaller sections. A safety catch would again be very necessary to prevent any movement of the lengths of metal.

Sheet material is best stored so that it rests against a wall in a near vertical position but with a rail (or rails) to keep it in position. Obviously it must be sited so that sheets can be removed or replaced easily and safely. Short ends of wood and metal are very difficult to store and yet they are much too valuable to throw away. Some teachers keep odds and ends of timber in a tea chest but it is impossible to get at anything buried beneath the top layer and eventually everything, in desperation and frustration, goes for firewood. A better system is to use a container consisting of wire cages giving a pigeon-hole system which enables short ends to be stored according to type and yet to be easily retrieved when needed. It may be possible to acquire surplus cages of the type used in Physical Education changing rooms to store gym

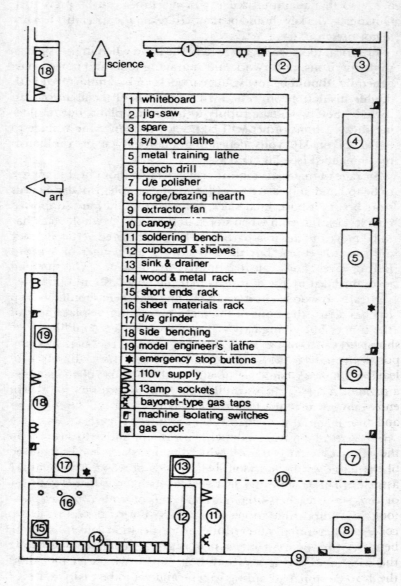

1	whiteboard
2	jig-saw
3	spare
4	s/b wood lathe
5	metal training lathe
6	bench drill
7	d/e polisher
8	forge/brazing hearth
9	extractor fan
10	canopy
11	soldering bench
12	cupboard & shelves
13	sink & drainer
14	wood & metal rack
15	short ends rack
16	sheet materials rack
17	d/e grinder
18	side benching
19	model engineer's lathe
✷	emergency stop buttons
⌇	110v supply
B	13amp sockets
⚥	bayonet-type gas taps
⌐	machine isolating switches
▣	gas cock

Figure 6.3 Middle school workshop layout

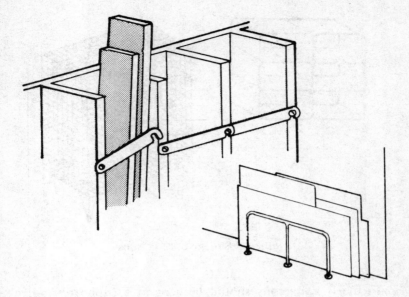

Figure 6.4 Storage of rod and sheet materials

shoes and clothing and to modify them to form a stack. Another possibility is to remove the tops from empty 5-litre oil cans, leaving the 'safe edge' intact, and to solder these together to form a pigeon-hole rack. This could be particularly suitable for short ends of metal and plastic rod and bar material (see Figures 6.4 and 6.5).

Storage of hand tools will be determined almost completely by their frequency of use. Tools which are needed by large numbers of pupils almost every lesson should certainly be in racks or trays associated with each work bench. The tools should fit into slots or recesses so that cutting edges are protected and so that each tool has a recognizable home. The tool handles should be colour-coded so that each set of tools is associated with a particular bench or vice, and in this way not only will it be easier to control the stock of tools but also the pupils will feel that the tools are their responsibility because they will know that they are shared by a limited number of other pupils.

Tools which are very rarely used or which are expensive or

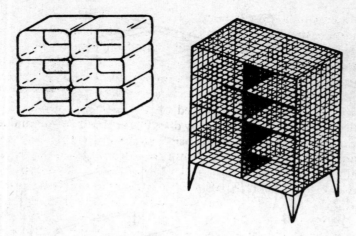

Figure 6.5 Storage of short ends

particularly dangerous should be kept in a cupboard and, if necessary, under lock and key. Other tools, however, should be on display in the workshop, preferably in open racks designed to display the tools attractively and safely. Wherever possible the rack should be sited to avoid unnecessary movement on the part of pupils and therefore should be sited nearest to the point of maximum use. No rack should ever be put behind a machine so that pupils have to reach over a potentially dangerous piece of equipment. Lathe tool racks, for example, should always be placed beyond the tailstock, and the practice of putting a sketch board or small blackboard behind the lathe bed is particularly dangerous.

If tools are racked in a wardrobe-type cupboard the backs of the doors should be used as well as the shelves to display the tools so that a 'shop window' display is effected when the room is in use. If the inside of the cupboard is painted white and the racks of tools labelled with Dymo tape or a similar form of lettering then the cupboard will become a useful and attractive teaching aid as well as a storage facility. Considerable thought must be given to the problem of designing effective tool racks. The tools must be held securely and safely, sharp cutting edges must be protected, the tools must be easily identified and they must be easily removed and replaced. Some tools, such as chisels, are best stored in a free standing rack which is kept in the cupboard when not needed but which can be carried round from bench to bench by a monitor to remove the need for pupils to carry individual tools around the area unprotected. In order to aid the correct return of tools to their proper places it is recommended that a silhouette of each tool should be displayed on the racks or shelves. These silhouettes could be created by drawing round each tool and then painting the shape but an easier and better method is to draw round the tool on a piece of thin coloured card and then to glue the cut-out shape onto the rack. Late on Friday afternoon a dark silhouette could be mistaken for the tool itself so it is better to use brightly coloured card rather than blue, brown or black.

As suggested earlier, it is advisable to have a spare set of common hand tools racked up in a drawer, tray or trolley so that these are always available for classroom use in cross-curricular activity. These tools should remain the responsibility of the specialist teacher and should be returned to the CDT area at regular intervals for checking, sharpening and other maintenance.

For most sketching and design work connected with problem-solving activity a set of simple A4-size clipboards made from a sheet of hardboard would be more than adequate for most pupils' use, but there should also be a couple of drawing-board stands with parallel motion, for use in special jobs or for use by exceptionally able students. A couple of units such as are supplied in the Thornton Desra range would be most suitable and they could be shared with the Art and Design area.

Sometimes the children's work needs to be painted and this creates problems in that a dust-free environment is needed and the models have to be kept away from exploring fingers while the paint dries. If there is room there is no better place for a painting table or bench than just inside the storage area provided that the pupils using it can be seen by the teacher from the remainder of the workshop.

If effective problem-solving activity is to take place then both motivation and sources of information will be needed in the CDT area. Every available wall surface should contain display boarding and the displays should be both eye-catching and relevant. A fair proportion of the display space should be devoted to the pupils' own work and there should be frequent changes of material. One particularly prominent section of pin-boarding could be reserved for displaying items connected with any group currently in the workshop area. In other words, it could be seen as an extension to the blackboard and therefore relevant to the groups' own work. If items on this section of board are kept to an absolute minimum, it is easy to change them between sessions and, of course, they will have greater impact.

A definite timetable should be established for the remaining display boards with, perhaps, one board changed every week (or every other week if a ten-day timetable is in operation) as if items are left on display too long they become part of the background scenery and completely lose their effectiveness. The display, too, must be planned and treated as an advertisement or poster with carefully arranged components and professional use of lettering and colour if it is to compete for attention with the impact of modern graphics and technical illustration to which the children are subjected daily. Used properly, the display boards in the area will be an enormous asset in aiding motivation and interest but to carry this motivation through to realization there must be access to knowledge, information, technical and technological concepts, and this pinpoints the necessity of providing adequate literary, visual and tactile resources. Books, booklets, manufacturer's brochures and other literature need to be readily available for use by the children in their problem-solving activity, but to be really effective there is a need for an index and some sort of retrieval system. The books are best housed on shelves behind glazed doors to keep them free from the dust which is inevitably generated in any workshop. Books on tools and techniques will, of course, be needed but the reference library should also include books on shape, form, surface pattern, texture, materials science, how things work, technological concepts, the built environment, electrical and electronics activity, structures, machines, pattern and colour. It may be possible to borrow books on an extended loan procedure from either the school's own library or from the Public Library service.

Magazines and journals are a useful source of information and these are best displayed on racks or reading stands. Earlier editions can be stored on shelves in a cupboard but, if information is to be retrieved from them, then some sort of cross-referencing system will be needed coupled with an index which can be added to regularly. Articles, magazine cuttings, trade leaflets and advertising literature are best stored in labelled box files but included in the cross-referenced system.

If there is to be a continuous rise in the standard of CDT activity in the middle school then it is essential that the pupils should benefit from the results and endeavours of the boys and girls who have preceded them. Examples of current and earlier work by pupils should be displayed, not only in the CDT area but also elsewhere in the school, perhaps in the entrance foyer or in the year base, and even greater benefits could be obtained by adding selected design briefs, relating to earlier work, to the reference library. This would enable pupils to start their researches and investigation at a more advanced stage than if they were starting from scratch (and, in fact, would reflect the story of human endeavour and achievement in that each 'generation' would stand on the shoulders of its 'forebears'). Once again these design briefs would need to be included in the cross-referenced index.

Three-dimensional artefacts and natural forms should also be available in the reference library as additional sources of knowledge and experience. During the middle school phase many pupils are still at Piaget's level of 'concrete' experience and need to be able to handle realia to make sense of their environment and the problem set. In addition to this aspect, however, all middle years pupils need to have frequent opportunity to handle things as at this stage the sense of touch is the most highly developed of the senses and this needs to be exploited.

One of the biggest headaches facing the middle years CDT teacher is that concerned with the storage of unfinished work. Small bits and pieces can be stored in plastic trays with a separate tray for each group housed in a purpose-made cabinet or stacked on the shelves in the side cupboards or under the workbenches. Rather larger models can present a bigger problem unless storage space is provided in the materials store area. One of the best solutions is to create a wall rack of open box sections, each at least 300mm square, by putting vertical spacers between wall mounted shelves over the painting table or cupboard in the store area.

Window sills, the tops of cupboards and side benching and,

above all, the floor, should be kept clean and tidy and free from the clutter caused by leaving bits and pieces lying around. Anything left on the floor could create a safety hazard by causing someone to trip or fall, and clutter elsewhere in the workshop collects dust as well as looking untidy and creating the impression of an unworkmanlike environment. If each group develops the habit of returning every tool and item of equipment to its proper place at the end of each lesson and if material stocks are tidied at the same time by another set of monitors then everywhere will remain neat and tidy with a minimum of supervision and effort.

Small stocks of consumable materials can be kept in a cupboard in the work area but the main stocks should remain in the storeroom. Flammable materials, of course, must be stored in appropriate containers to comply with fire and health and safety regulations, and in cases where larger stocks are held a proper steel cabinet or brick store must be used. If in doubt it is highly advisable to consult the CDT Inspector or Advisory Officer to avoid the risk of committing an offence.

Finally, remember the motto, 'a place for everything and everything in its place', and you cannot go far wrong as far as tidiness and good housekeeping are concerned.

Chapter Seven

RESOURCES FOR LEARNING

The use of computer and information technology and other advanced technologies in control, simulation and data storage and retrieval is becoming increasingly important in our society. This fact should be reflected in the use of computer and information technology across the school curriculum.

(National Curriculum Report: Science for ages 5 to 16: DES 1988)

The description of technology as given earlier in this book is that it is connected with the purposeful use of knowledge, materials and resources to satisfy a human need. Similarly it was stated that design is concerned with aesthetic considerations and an under-standing of the functional 'rightness' or fitness for purpose of man-made objects. Obviously children in school cannot relate to these aspects in their work in CDT unless the learning environment is in tune with their needs with the teacher acting as motivator, guide and tutor and with appropriate resources to hand, particularly during the exploratory and divergent stage of the problem-solving process. In order for this to be possible, not only must appropriate and selected resource material be to hand, but the classroom or workshop itself must be seen as an import-ant part of the learning provision.

In establishing the range of resources needed for work in CDT (or any other subject activity for that matter), the only really effective way to proceed is for as many teachers as possible to contribute material under the guidance of a 'resource co-ordi-nator'. At present, in far too many schools individual teachers have their own visual aids and other teaching material which is never made available for use by other colleagues. This certainly results in duplication of effort but also in inefficient use of total school resources as far as individual pupils are concerned. Obviously teachers should have first call on material which they have produced for use with their own class or classes, but surely this same material could be made available for use by teacher colleagues at other times.

Before setting up a resource facility the first requirement is to establish the type of recall system which is to be used. Some teachers prefer to use the school or class library as a starting point using the same colour coding or Dewey type reference system in operation for the library but extending it to cover pictures and other printed reference material. Others find it more efficient to set up an appropriate classification system for the recall of the collected resource material and to cross-reference the library system into it. The important point is that the system adopted must be logical, sensible, easily understood and appropriate for the age and ability range of the pupils who will use it. For example, a green label could indicate that the book or box contains information on buildings. A gold dot on the green label could indicate more precisely that the contents covered church buildings. Further coding could relate to specific historical periods such as Saxon, twentieth century, and so on, the degree of subdivision depending on the age and ability levels of the pupils who will refer to it. Many teachers, however, prefer not to use colour coding because of the prevalence of colour blindness, particularly in boys, and prefer to use letter or figure codes instead.

Figure 7.1 (Resource indexing) shows how sections and subclassifications could be used to develop a recall system. In the extract illustrated an example could be that pictures and other information about British steam trains would be found by looking up section B (Transport) and then turning to material indexed under 81 (Trains), 1 (Steam) and 18 (Great Britain). To narrow the selection down even further the classifications 113 to 119 would lead us to specific historical periods in the development of the railway system.

The system can be as simple or as detailed as deemed appropriate by the organizer, and with very young children different shapes or even basic pictures of the class of objects could form the index system. Naturally the classification for younger or less able

A MACHINES	B TRANSPORT	C BUILDINGS
D POWER	E MATERIALS	F OPTICAL
G FLIGHT	H UTENSILS	J FURNITURE, etc.

1 Steam	41 Slide Proj.	81 Trains
2 Coal	42 Cine Proj.	82 Cable Cars
3 Electricity	43 Overhead Proj.	83 Lifts
4 Gas	44 Periscope	84 Airliners
5 Water	45 Kaleidoscope	85 Helicopters
6 Wave	46 Mirror	86 Churches
7 Solar	47 LED	87 Halls
8 Muscle	48 LCD	88 Hotels
9 Turbine	49 Prisms	89 Houses
10 Battery	50 Ores	90 Bungalows
11 Mechanical	51 Forests	91 Castles
12 Atomic	52 Woods	92 Markets
13 Gravitational	53 Man-made	93 Farms
14 Chemical	54 Ferrous	94 Factories
15 Diesel	55 Non-ferrous	95 Shops
16 Petrol	56 Thermoplastic	96 Offices
17 LPG	57 Thermosetting	97 Stations
18 Great Britain	58 Resins	98 Tower Blocks
19 France	59 Adhesives	99 Generators
20 Germany	60 Paper & Card	100 Water Works
21 Holland	61 Concrete	101 Markets
22 Sweden	62 Tubes	102 Colleges
23 Norway	63 Laminates	103 Stone Age
24 Spain	64 Brick	104 Bronze Age
25 Italy	65 Stone	105 Iron Age
26 Japan	66 Glass	106 Greek
27 Russia	67 Textiles	107 Egyptian
28 China	68 Plaster	108 Roman
29 N. America	69 Cars	109 Norman
30 S. America	70 Vans	110 Saxon
31 Africa	71 Lorries	111 Medieval
32 Australia	72 HGVs	112 Elizabethan
33 Brazil	73 Trams	113 Georgian
34 New Zealand	74 Trolleys	114 Victorian
35 Telescope	75 Taxis	115 1900–14
36 Binoculars	76 Motor Cycles	116 1915–40
37 Microscope	77 Omnibus	117 1941–60
38 Spectacles	78 Liners	118 1961–80
39 Magnifier	79 Ferries	119 Post–1980
40 Camera	80 Underground	120 Satellites

Figure 7.1 Resource indexing

children would need to be simpler than that for older or brighter groups, but if the system is devised sensibly the search can be undertaken step by step to find the necessary information. There are, of course, many systems in use for resource indexing and cataloguing, such as card indexing and the use of light boxes (Optical Coincidence Co-ordinate Indexing – OCCI), but most of these earlier methods have been superseded by microfiche or computer programs to obtain easier retrieval.

On many occasions it would be necessary for the teacher to extract the required materials needed from the resources library so that they could be immediately available in the classroom or workshop for the needs of the problem-solving or project activity.

In some schools a central resource area is created, perhaps based on the library, and all material is concentrated at that point. If a full-time librarian is employed then, with careful organization, this system can work, but in most schools it is far more effective for the resource material to remain in individual classrooms and specialist areas but for everything to be catalogued on a central list or index system. This system would need an additional item of information indicating the room where the material is to be found. No matter what system is used, the building up of a comprehensive index is a long, slow process. However, electronic methods enable additions and updating to be effected much more easily than with earlier systems. Schools with easy access to a computer have an advantage, of course, in that a program could be devised to contain all the information needed for easy retrieval. The alternative is to use a card index system perhaps related to a heading summary on a central catalogue or list which could be displayed in each room.

Apart from the need to cross-reference books and their contents into the central index, what other types of material could be useful for CDT reference purposes? Any printed item or piece of realia which is concerned with materials, processes or properties connected with the needs of mankind would be relevant. The list would therefore include magazine and newspaper articles, cuttings and pictures; articles and cuttings from educational journals and home magazines; postcards; advertisements; brochures and trade or educational literature from industrial sources, commerce and government departments; slides; audio and video tapes; films; OHP overlays and transparencies; teachers' work sheets and work cards; visual aids; posters; wall charts; materials; samples; constructional kits; examples of realia (for example, gears, nuts and bolts, clock parts, bicycle parts and accessories, labelled samples of timbers, metal and plastics, and so on).

Newspaper and magazine cuttings and pictures are best mounted on sheets of paper or thin card to make them more durable. Manilla folders or envelopes can be used to house print material or, better still, shallow boxes such as shirt boxes or plastic storage trays can be used. The important aspects are that the containers should be of a convenient size, stack or store easily, be easy to get at and that they are able to cope with further items as additions and deletions are made.

Even with someone acting as co-ordinator, and with a number of teachers contributing to the system and helping with standardized indexing, it can take several terms to get the resource system fully operational and established. The problem subsequently is to maintain the momentum and enthusiasm so that the contents can be regularly updated, obsolete material removed, damaged material repaired and so on. If cuttings are slipped into loose-leaf folders or manilla envelopes under broad general headings as soon as they are received then it is not too difficult or time consuming subsequently to mount and catalogue them and to slot them into the system. A pile of unsorted material, however, is unlikely ever to be transformed into a usable system.

In many schools the general classification is done by the teacher receiving the material but the actual entry into the resource system is carried out by ancillary staff or by parent helpers.

Children's own work, especially work connected with projects or with fields of interest, could be added to the system. This can have the advantage of enabling each new piece of work to be extended beyond the level of previous work thereby leading to real progression and ever higher standards and levels of achievement.

Obviously, the library is an important part of a resource system and, as well as helping children to find appropriate information by use of a coding system, there must also be regular help and

121

guidance in using indexed books for reference purposes where the same topic is looked up in several books, the information that is gleaned being subsequently précised and collated. The County or Borough Library Service will normally lend reference and other books and materials to schools for an extended period and this facility should be used to supplement the school's own library. By planning ahead it is normally possible for arrangements to be made for the County or Borough Library to send a package of books on particular topics or fields of interest and this can be invaluable in subject areas such as CDT where the school's own stock of books may not be very comprehensive. Similarly, many museums and art galleries operate a loan service for schools and this facility can and should be exploited to add to the school's own resources. As mentioned earlier it is usually advisable for the teacher to draw on appropriate resource material to anticipate the needs of the class or group and to prevent wasted time and the creation of supervisory difficulties which could result if pupils have to move around the school to find and collect their own resource material. In the secondary school, however, it should be possible to rearrange facilities to create a resource area in the CDT department itself, ideally to be shared with both the Art and Craft and Home Economics departments. This area would contain all the indexed and catalogued material appropriate to the three departments together with an overhead projector and a simple back-projection screen slide projector, so that individuals or small groups of pupils could carry out their graphic and problem-solving activity in a clean, dust free, and relatively quiet environment with print and non-print resources to hand. If the shared resources area has glazed partitions then there are few supervisory problems and the room can double as a tutorial room or even a technology room for groups of twelve to fifteen pupils.

Finally, mention must be made of the classroom or workshop as a resource in itself. The room must be well organized with materials, equipment, tools and guidance readily available, and if the tools are properly racked and labelled they become part of the resource. The display boarding, too, must contain attractive and well-presented visual items relating to the work in hand so that they are informative, as well as serving to motivate the pupils. Too many classrooms and workshops are either dull, boring places, or else they are so cluttered with paraphernalia and work in progress and with out-of-date material on the display boards that they make no impact on the child. The classroom or workshop should be seen as an important teaching aid and the most valuable resource of all and therefore presented accordingly.

Chapter Eight

ASSESSMENT AND EVALUATION

The purpose of assessment is to show what a pupil has learned and mastered, so as to inform decisions about the next steps, and to enable teachers and parents to ensure that he or she is making adequate progress . . . assessment should be seen as part of the teaching process and made in such a way as to give valid information.

(National Curriculum Report: Science for ages 5 to 16: DES 1988)

A society which cannot, for whatever reason, recognise and affirm the necessity, value and virtue of the activity by which it principally earns its living, faces a major dilemma. It is unable to say YES to its own future because it does not say YES to the activity on which that future depends.

(Kenneth Adams, 1977)

The earlier parts of this book explored the theme that our priorities regarding the relative importance of subjects in the school curriculum could be ill-conceived and no longer relevant to the present and future needs of our pupils in view of the rapidly changing and largely unpredictable nature of our society. It is accepted, of course, that literacy and numeracy underpin almost all other activities and that these should have pride of place in the timetable provided that they are not seen as isolated activities divorced from other curriculum areas and the argument has already been developed that CDT could be an ideal vehicle for the application of communicative and mathematical skills and as a motivator and source of incentive for these basic subject areas. However, the fact remains that when judgements are made about a pupil's performance or level of ability then enormous emphasis is normally placed on language and number skills and attainments while other curriculum areas such as CDT, Art and Design, Music and Physical Education may be virtually discounted other than for a passing reference in a record book or report sheet. Are these other areas really of so little importance and, if so, should we continue to include them in the school day?

Although in the larger society the actor, musician, painter, sculptor, ice-skater, athlete and sportsman receives varying degrees of recognition according to luck, ability or media promotion, it is extremely rare for any engineers, designers or craftsmen to be accorded the respect which their importance to the twentieth century society demands, and this prejudice may have some connection with the decline of the manufacturing and production industries in Great Britain in the second half of the century.

In school, most teachers would probably claim that they are endeavouring to educate the whole child to enable that child to take his or her place in society, to lead a full and satisfying life and to contribute and benefit according to individual ability, but when asked to assess the child's ability, perhaps for banding or streaming purposes, the important criteria in the eyes of the school are often concerned solely with literacy and numeracy. The skills and attitudes relating to spatial concepts, mechanical functions, manipulation of materials, aesthetic sensitivity and applications of energy are unseen and unrecognized. Surely this must not be allowed to continue?

Of course, there are problems. What does the teacher look for when assessing, say, a working model or a piece of craftwork? How can personal prejudice, subjective judgements and attitudes to fashion and style be replaced by appropriate assessment objectives? Is the aim to assess the piece of work produced or to evaluate the effect which the process has had on the development of the child? Should the nature of the work set be such that it lends itself to the process of testing? These and many more problems come to mind, exaggerated by the fact that very few teachers have any real experience in assessing practical subject areas and that there are very few examples or accepted and proven guidelines which the teacher can apply.

It is understood that the Assessment of Performance Unit (APU) is continuing to investigate the problems of 'Understanding Design and Technology' and a useful discussion paper bearing that title and circulated for comment during 1981 and 1982 is still very relevant. Further statements from this valuable and respected source could be extremely helpful to teachers concerned with the wider assessment of pupils in the 5–13 age range, particularly in that it can establish norms by reporting across large numbers and that it can advise on *what* might be assessed as well as *how* it may be assessed.

One of the problems is that CDT does not lend itself to testing under examination conditions and it would be very wrong to distort the subject to enable it to fit into an artificial examination pattern. Indeed, there are numerous examples in middle and secondary schools where the year's work in CDT has been concerned largely with the design/make/evaluate process but where the end-of-year examination consists of a 90-minute written paper requiring memory and written communication skills rather than those skills which have been dominant in the course work. Fortunately, various assessment techniques can be applied to this subject area including a major emphasis on the course work itself and, because of the nature of the problem-solving/design process, self-evaluation should be seen as an important element.

It must not be assumed that everything can be quantified nor that every teacher has the ability to assess every facet of a child's development. If this were so then school life would consist of one long test (or perhaps it does for some pupils!) and the present situation where external examinations appear to control the school curriculum could deteriorate still further into a rigid and inflexible timetable with no room to manoeuvre or to exploit situations and interests as they occur. The introduction of the GCSE examination in 'CDT: Design Realisation', 'CDT: Technology' and 'CDT: Design and Communication', with the associated project work, has done much to liberate the curriculum for the older comprehensive school pupil, and there should be more freedom from examination constraints for the younger comprehensive and middle years pupils as well. Perhaps now, for the first time in CDT, the examination will be determined by the curriculum rather than vice versa.

In the brochure, CDT – *A Curriculum Statement for the 11–16*

Age Group, published on behalf of a group of HMI in 1983, it was stated that:

> It is quality of the pupil's unified experience of designing, planning, making, testing and evaluating that is of fundamental importance·rather than any ability he or she may acquire in a specific competency.

This surely means that CDT assessments should be primarily concerned with this unified experience, but what are the aspects which should, or could, be assessed or quantified? Certainly those cognitive and manipulative skills concerned with the processes of designing and making should take pride of place, but any assessment on these lines must recognize the importance of aesthetic sensitivity in the development of the child. Because problem-solving is a unified, cyclical process, perhaps it would be wrong to allot objective marks to each section or stage of the process in which case it may be better to apply some kind of subjective assessment to the overall process. If this is done how would the pupil who has produced a mundane but well-made and functional item be assessed in comparison with the equally skilled pupil who has explored an exciting line of development but whose solution fails in some way or other to do what is expected of it? Should we, in fact, make assessments based on expectations or should they be based on established norms or even on potential?

This line of questioning could continue but perhaps the reader could use these points as the trigger for lively staffroom discussion. However, this book aims at giving pragmatic help and advice, so perhaps the first thing which should be done when drawing up an assessment policy is to reconsider the aims of teaching CDT and to analyse the real objectives. This must include a self-evaluation policy on the part of the teacher starting with an analysis of the effectiveness of our own teaching set against the original aims and objectives. Some of those aims would be general in educational terms, others specific to certain aspects of CDT but, taken together, they could be seen as our attempt to educate the whole child as stated earlier in this chapter. It is then necessary to look at the educational, social and moral content of the activities and experiences encompassed within the CDT package and to decide which of them are relevant and worthy of assessment as a record of the child's current development and abilities.

Inevitably this leads to the need to consider the value of some form of profile assessment as compared with bare marks or grades, perhaps averaged from quite disparate individual assessments. For example, in the unlikely event of a pupil receiving an 'A' for aesthetic sensitivity, 'B' for ideas, 'C' for effort, 'D' for technological knowledge and 'E' for tool skill, would a final average mark of 'C' mean anything of value when it was entered on the pupil's annual report?

One big advantage of the profile assessment method is that the profile can be related to definite stages in the pupil's life and can therefore refer specifically to those aspects of development and achievement which are brought about by the current aims and objectives for that individual or group. Another is that a profile consists of observations made by the teacher, rather than opinions, and therefore it can be seen to be relatively objective in nature. A third advantage is that it is diagnostic and can be used as the basis for subsequent teaching programmes whereas the conventional report sheet is normally only congratulatory or condemnatory and of little subsequent value to pupil, parent or teacher.

What goes on a pupil's profile sheet would depend on the phase of development, the role of CDT in the curriculum, the nature of class or group organization and the interests or expertise of the teacher, but the following suggestions may help the reader to evolve experimental and appropriate models.

Obviously, in such a diverse curriculum area as CDT, where pupils are involved progressively in an ever-widening range of activities and experiences, and where tool skills and techniques can be developed as physical strength and manual dexterity improve, it is difficult to list a finite set of aims and objectives and therefore it is equally difficult to define precise assessment objectives for different age and ability level groups. Nevertheless, it may be useful to list some of the activities and experiences which should be an integral part of CDT work and which apply generally throughout most of the early and middle years phases of education (see also Figures 8.1, 8.2, 8.3).

ASPECT	ACHIEVEMENT
Has the ability to recognise wood, metal, plastic, clay.	✓
Recognizes the 'fluid' property of dry sand.	✓
Has used damp sand in constructive play.	✓
Can construct using bolt-together plastic kits.	
Can construct using click-together plastic kits.	✓
Can build by bonding wooden bricks.	
Can construct and dismantle using Lego.	
Can create wheeled models using Lego.	
Can describe the stages in building a wooden block archway.	
Can use a gent's saw.	
Can use hammer and pegs.	✓
Can use hammer and nails.	✓
Can use hard, soft, stiff, bendy, floppy in context.	✓
Can use rough, shiny, smooth, dull in context.	✓
Can name saw, hammer, nail, pencil.	✓
Can work with a group.	
Prefers to work alone.	✓
Respects other children's work.	✓
Shows imagination.	
(NB These are given as examples only and do not constitute a comprehensive profile.)	

Figure 8.1 Part of a possible profile sheet for a 6-year-old pupil relating to CDT

ASPECT	ACHIEVEMENT
Can use correctly and name gent's saw, hammer, pliers, screwdriver, bradawl, pincers, file, surform tool, junior hacksaw, bench hook, vice.	✓
Can use nails, screws, nuts and bolts.	✓
Can use soft solder.	✓
Can use PVA adhesive.	✓
Can manipulate acrylic resin sheet.	
Can shape wood by addition or reduction.	✓
Can 'finish' to a good standard.	
Can apply paint or shellac polish.	✓
Can design and make wheeled Lego vehicles.	✓
Can use Meccano or Stokys kits.	✓
Can apply principles of pulleys and levers.	✓
Understands role of gears and gearing.	
Can deduce cause and effect in connection with mechanisms.	✓
Can solve simple design problems using card, wood, metal and plastics.	✓
Can design and make a powered vehicle.	
Can measure and record in metric units.	✓
Understands principles of scale.	
Can communicate ideas by drawing.	✓
Can compile an effective design brief.	✓
Can co-operate with other pupils in the group.	✓
(NB These are given as examples only and do not constitute a comprehensive profile.)	

Figure 8.2 Part of a possible profile sheet for a 9-year-old pupil relating to CDT

ASPECT	ACHIEVEMENT
Can design, make and evaluate simple artefacts.	✓
Can design, make and evaluate simple energy-using devices.	✓
Can accurately shape and form wood, metal and plastic sheet.	✓
Understands orthographic projections.	✓
Can produce effective working drawings.	✓
Can communicate by free-hand drawing.	✓
Can communicate ideas verbally using correct terminology.	
Is proficient with common hand tools.	
Has a good understanding of mechanical control.	
Has a good understanding of electrical control.	✓
Has used microcomputer control.	
Has a good understanding of the social implications of technology.	
Can use the bench drill safely.	
Can use the training lathe safely.	✓
Has a sound knowledge of the main properties of common materials.	✓
Is an excellent/good/average/poor contributor to group activity.	Av
Can relate today's achievements to historical developments.	
Is developing an aesthetic awareness.	
Shows initiative.	
Tackles problems imaginatively.	
(NB These are given as examples only and do not constitute a comprehensive profile.)	

Figure 8.3 Part of a possible profile sheet for a 12-year-old pupil relating to CDT

Through the operation of the Craft, Design and Technology syllabus all pupils should have regular experience in the areas listed in the following checklist:

- Frequent practice in analysing suitable design problems leading to design/make/evaluate activities.
- Opportunities to use their own ideas and to exercise their imaginations.
- Research into relevant ideas and resources.
- Selection and development of appropriate solutions.
- Encouragement to persevere towards a chosen end.
- Use of drawings and diagrams to convey ideas and information.
- Use of simple plans and elevations to communicate precise information.
- Working from plans and diagrams which they have prepared.
- Use of models and mock-ups as prototypes to confirm ideas.
- Use of scale in transferring from drawing to construction.
- Use of symbols in visual communication.
- Use of drawings to demonstrate progression.
- Use of drawings to demonstrate how things work.
- Controlled use of a range of durable materials.
- Practice in manipulating, modifying, joining and assembling durable materials.
- Safe usage of a range of hand tools with correct acquisition of dexterity and skill.
- Knowledge of properties of common durable materials.
- Development of sensitivity towards materials and processes.
- Opportunities to test properties of materials.
- Progressive additions to vocabulary, especially relating to materials, techniques, aesthetic experiences and technological awareness.
- Practice in verbal communication relating to CDT activities and experiences.
- Sensory experiences in handling and working with materials, tools and processes.
- Evaluation of man-made artefacts; their suitability, function and form.
- Experiences associated with applications of different forms of energy.
- Model-making using different mechanical and electrical

control techniques.
- Familiarity with the world of technology and an increasing technological awareness.
- Technological experiences concerned with an interaction with the environment.
- Practice in applying discriminatory skills.
- Social awareness about the effects which technology can have on modes and standards of life.
- Drawing from direct observation of man-made artefacts.
- Experiences in solving problems by making things that work.
- Acceptance of the need for a high standard of finish and for correct craft skills.
- Development of a sense of satisfaction based on achievement.
- Experience of working in groups towards a common goal.
- Experience in relating to others where each member of the group contributes a particular skill or attribute.
- Respect for other pupils' work and for the environment and the development of an associated sense of responsibility.
- Development of an awareness of health and safety needs.

These, or similar activities and experiences, are certainly the *raison d'être* for the inclusion of CDT in the early and middle years' curriculum, and therefore they would appear to be the aspects which should be assessed and evaluated when attainments and capabilities are under review. Some of them defy analysis if seen as separate categories but, as part of an overall activity or experience, they can be included by continuously assessing the pupils' coursework, provided that it is the pupil who is being assessed rather than the artefact that is under construction.

The introduction of the National Curriculum with its programmes of study and attainment targets requires assessment and evaluation at the ages of 7, 11, 14 and 16 years, and this will alter the whole pattern of teacher/pupil assessment as the results will be moderated externally from the school. Most of the assessment will be carried out as part of normal classroom activity but there will almost certainly also be nationally prescribed tests for all pupils to supplement the teachers' internal assessments. Secondary teachers who are familiar with the patterns of the external examination system will readily adapt to the new approaches, but most teachers of younger children will need to avail themselves of INSET help if their assessments are to serve the purpose for which they are intended.

Earlier in this chapter mention was made of the importance of self-evaluation on the part of the pupil and this should be seen as an integral part of the design process. The pupil should be expected, within reason, not only to evaluate the effectiveness of the end product, or prototype, but also to try to assess what he or she has gained from the activity. The place for this is in the design brief which each child should compile, each brief containing not only notes and drawings made during the investigatory stages but also notes and comments entered while the work is in process and at the conclusion of the activity, these notes including reference to successes and failures, personal feelings about the work, comments on attitudes and satisfactions and, finally, a few words about fitness for purpose and what the child has gained, in personal terms, from the work. Of course, younger and less able children would have to carry out this evaluation verbally, perhaps in a question and answer sequence with the teacher, perhaps in discussion with others in the group, but it is very important that all children are encouraged to evaluate their own work as this has the added advantage of enabling children to set their own high standards rather than having standards imposed from outside.

This chapter has set more questions than it has answered but this is inevitable where we are trying to decide how to assess such a wide ranging activity. In many instances there are no 'correct' answers to the problems under investigation by the pupils and certainly there is need for major research into assessment and evaluation techniques in this increasingly important subject area. What must be remembered is that assessment needs to reflect the unified approach which is inherent in CDT activity and that any assessment of specific skills and activities is only of value if seen as part of the greater whole. Also, that equal weighting must be given to this area of experience, alongside the other areas of experience making up the total curriculum, when assessing the child's abilities and potential. If justification is needed for this to happen then perhaps a quote from an internal document circulated as a curriculum statement to members of the Association of Advisers in Craft, Design and Technology in 1980 may be of importance:

The educational experience of Craft, Design and Technology is a preparation, and in parts a rehearsal, for how the future generation will order their environment. It is the beginning and development of a set of enabling skills and knowledge to increase human potential.

Chapter Nine

HEALTH AND SAFETY CONSIDERATIONS

We were also asked to consider aspects of health and safety. We took the view that these matters should feature directly in a number of themes and . . . through the development of relevant skills and attitudes.

(National Curriculum Report: Science for ages 5 to 16: DES 1988)

Teachers have always held a responsibility for the safety of the children in their care, but recent legislation has made everyone increasingly aware of the need for safe working conditions, not only in the place of employment, but also in the home and on the roads. HM Inspectors of Factories have long held a responsibility for the standards of safety in industry, but since the publication of the Health and Safety at Work, etc. Act 1974, those inspectors have a right of entry to schools and they now work closely with HM Inspectors of Education to ensure that the work in schools is adequately safeguarded. The Act places a general responsibility on employers, employees and others to ensure that working conditions are safe and, although the employer has a responsibility to provide appropriate courses of training, the employee has a duty to comply with any regulations regarding safety and to ensure that no unsafe practices are used or unsafe conditions created. Children are not regarded as employees but they are still covered by the Act as are visitors or others on school premises.

In simple, basic terms, the teacher is expected to give reasonable supervision to all children in his/her care and to act at all times as a reasonable parent would act. The important word here is 'reasonable', as no court would expect a teacher to do the impossible unless the unsafe conditions were created by the teacher in the first place. There is also a responsibility placed on the teacher to ensure that appropriate training has been received before the machine or technique is used and the teacher would certainly be expected to be aware of any safety hazards associated with particular processes or activities.

In the school without specialist workshop facilities the main responsibility would be to ensure that the children could work safely in uncluttered surroundings and that the use of any potentially dangerous tools or equipment was correctly taught and demonstrated and properly supervised with the children being fully aware of the dangers and how to avoid them by safe working practices.

In the classroom, as has been stated in the chapter on classroom organization, it is sensible to create a corner of the room with improved circulation space where CDT activity can take place. Not only does this create safer conditions but it also makes supervision so much easier. No tools with cutting edges should be permitted unless their use has been carefully and recently demonstrated, their safe use is revised on each and every occasion the tools are introduced and close supervision of the work in progress is possible. No tools with sharp cutting edges should ever be carried around the classroom unless they are in a properly racked tray or container with the sharp edges protected from accidental contact. Remember that a pair of scissors with sharp points can be just as dangerous as a chisel or a knife.

In large rural authorities, where local courses in tool usage may not be available, it is essential that primary teachers should make contact with their nearest comprehensive or high school and arrange for tuition in the safe and correct use of the basic hand tools from the staff of the CDT department, before introducing work with wood, metal or plastics in the classroom.

Every teacher should be familiar with the DES 'Safety Series', and particularly with *Safety in Practical Studies*, which was published by HMSO in 1981.

In schools with specialist workshop facilities, additional responsibilities rest with the teacher because of the wider range of equipment and processes available. CDT teachers working in the middle or secondary phases of education should be familiar with the DES publication *Safety in Practical Studies*, but they should also work within the guidelines laid down in the British Standards Institution publication BS 4163:1975 (or latest revision) *Health and Safety in School Workshops*. This latter publication is regularly revised and updated and every school should have a copy of the latest edition available. As with the primary school, later phases of education should ensure that in CDT all pupils work responsibly and with a clear understanding of safe working practices and of safety hazards. Careful organization of the work pattern is essential if adequate supervision is to be maintained throughout the lesson period with a number of different activities in progress. Protective clothing should be worn appropriate to the work being done and for most timetable activity this would consist of an apron and strong shoes as being the only provision outside normal school clothing, but when machines are being used, or if the work results in dust, fumes or particles being created, then goggles or safety spectacles of the correct British Standards specification must be worn. Similarly, goggles must be worn if solvents or plastic hardeners are being used or are being mixed, but this latter process should be carried out only by the teacher.

All machines should be fitted with guards and these must always be in place before the machine is switched on. Any pupil using a machine must have received appropriate training in the use of that machine and the teacher must be properly trained in the safe use of the machine and be able to supervise its safe use. It is advisable for pupils using machines to ensure that aprons are tied behind the body so that there are no loose tie strings hanging at the front, the strap on goggles should be safely tucked in so that there are no dangling ends and long hair should be tied back so that it cannot be caught up in any moving part on the machine.

The electricity supply to powered tools should be under the direct control of the teacher with a key-operated switch being used to isolate the supply to prevent unauthorized or unsupervised use of any machine except under the teacher's guidance and control.

The floor in the working area must be kept clear of materials and clutter at all times as many accidents are caused by pupils tripping over obstacles left around benches. Perhaps one of the biggest culprits would be the child's duffle bag and provision must always be made in a workshop area for the pupils to store their bags and other paraphernalia out of the way for the duration of the lesson. The floor in a CDT area is liable to become burnished if many wood shavings are created and if this happens steps must be taken to ensure that correct non-slip polishes are used by the cleaning staff to reduce the risk of slipping on a polished floor. A temporary cure can be effected by sprinkling a spoonful of rosin on to the affected area but it is far better to look for a permanent cure. If all else fails then adhesive pads can be purchased which have a slip-proof surface and these can be applied adjacent to workbenches and machines, but particular care must then be taken to ensure that the pad is firmly attached to the floor as a curled edge could create yet another hazard.

With such a wide range of materials available for use in CDT, care must be taken to ensure that no toxic fumes or by-products are created, especially if the material has not been purchased as 'non-toxic' from a reputable educational supplier. CDT teachers are well known for their ability to scrounge surplus materials from industry and elsewhere and this attribute is to be encouraged especially during times of financial stringency, but if there is any doubt at all about safety of a material then it should not be used. If plastics are being used then good ventilation is essential and it is advisable for the pupils carrying out the work to be situated near to an open window. Goggles must always be worn if there is any danger of fumes, liquids or dust particles entering the eyes and if dust is being created, as perhaps in lapidary work, then a mouth and nose dust mask should also be worn. Any mixing of hardeners with resins or similar processes should always be carried out by the teacher, away from where pupils are working, and of course the teacher must wear suitable protection. Schools envisaging work with plastics would be well advised to obtain a copy of the publication *Plastics in Schools: Safety and Hazards* from the Educational Officer, Plastics Institute, 11 Hobart Place, London SW1W 0HL, before commencing work.

Finally, workshops should be warm and well lit to create the safest possible working conditions. Pupils cannot work safely if they are cold or over-tired, or if lighting conditions are poor. Probably the most vulnerable time is the last hour on a Friday afternoon in November!

Chapter Ten

THE WAY AHEAD

In our desire to raise standards we have been conscious of the
need to build on the good work which is already going on in
many of our schools.

(National Curriculum Report: Science for ages 5 to 16: DES
1988)

Since the publication of the first edition of this book in 1985
considerable improvements have occurred in the status of CDT
in schools, and technology has at last been recognized as an
essential component of the curriculum, not least in the current
legislation for a National Curriculum in England and Wales.
Advances in training facilities for teachers have contributed to
improvements in teacher expertise in the area of CDT, especially
in the primary phase, and this is resulting in a greater confidence
amongst primary school teachers to extend the range and
standard of CDT activities with their pupils. Primary teachers,
too, are now identifying the technological content of the cross-
curricular work which has been such an important component of
their classroom activity and, in some cases, their work could be
said to be more closely related to the needs of CDT than that of
some of their more specialist secondary colleagues.

During this same period the general public has become much
more aware of the need to protect and conserve the environment
and of the role of technological activity in improving, and
sometimes destroying, that same environment. There also
appears to be a recognition that, if the United Kingdom is to
survive as a trading nation within the European Economic
Community (EEC) and across the world, then it becomes essential that our educational pattern produces young men and

women who are adaptable and enterprising and who can contribute to the needs of a twenty-first century society.

As every reader of this book will know, current legislation requires all pupils in England and Wales between the ages of 5 and 16 to follow a National Curriculum incorporating the 'core subjects' of English, Mathematics and Science, and a number of 'foundation subjects' including Design and Technology. Each of these subjects will have nationally determined, age-related, attainment targets and programmes of study and a national programme of assessment and testing at the ages of 7, 11, 14 and 16 years. This means that, in the very near future and for the first time ever, all pupils in England and Wales will have to receive a broad education throughout the period of compulsory education and they will not be able to abandon certain curriculum areas at the end of their third year in secondary school as was the case under former 'option' systems. It also means, of course, that the traditional primary school curriculum will have to change as all pupils must experience Design and Technology as an activity throughout the school.

Mention has been made of the recent improvement in the CDT expertise of many primary school teachers but, unfortunately, not many infant phase teachers have received adequate training in CDT teaching methods and this must be remedied urgently. Boys and girls entering school at the age of 5 years in September 1989 will be required to follow the National Curriculum in English, Mathematics and Science from the start of their schooling, and it appears that there will be 'unreported' assessment of their abilities in those core subjects when they reach the age of 7 in the summer of 1991. Pupils entering school in the autumn of 1990 will also be subject to 'unreported' assessment of Technology in the summer of 1992. It follows that all primary school teachers must avail themselves of every training opportunity and, in addition to the training initiatives of the local education authorities (LEAs), there is now in operation a centrally-funded programme in which, over a two-year period, two teachers from each primary phase school in Science, Mathematics, English and Technology will attend a two-day course of training supplemented by the work of teams of advisory teachers working alongside staff in their own schools. The courses are based on good primary practice, extending the existing curriculum but using national programmes of study for advice and guidance. There will be no need to change good primary practice and methodology (indeed, secondary school teachers could benefit by adopting certain primary phase methods) but curriculum and assessment structures may be improved to cater for the wider curriculum content and for greater expectations.

A DES press release in the spring of 1988 announced the setting up of a National Curriculum Working Group on Design and Technology by the Education Secretary, Kenneth Baker, stating that:

> Design and Technology are vital areas of the curriculum. They are of great significance for the economic well-being of this country. I believe it is essential that we press ahead quickly in establishing them within the national curriculum.
>
> My Rt Hon. Friend the Secretary of State for Wales and I are from today establishing a Working Group on Design and Technology. The group will advise on attainment targets and programmes of study for technology within the national curriculum for secondary school pupils. The Science Working Group which we appointed last summer is making recommendations about technology for primary school pupils.
>
> The Design and Technology Working Group will also advise on attainment targets and programmes of study for information technology, and on a framework for design across foundation subjects, for all pupils of compulsory school age.
>
> This group will start work as soon as possible. We are asking it to give interim advice by 31st October [1988] and final advice by 30th April 1989.

(DES press release dated 29 April 1988)

The 'Terms of Reference' for the Working Group state that:

> The Education Reform Bill . . . provides for the establishment of a National Curriculum of core and other foundation subjects for pupils of compulsory school age in England and Wales. For most subjects, including technology, the Government wishes to establish clear objectives – attainment targets – for the knowledge, skills and understanding which pupils of different abilities and maturities should be expected to have

acquired by the end of the academic year in which they reach the ages of 7, 11, 14 and 16; and to promote them, programmes of study describing the content, skills and processes which need to be covered during each stage of compulsory education. Taken together, the attainment targets and programmes of study will provide the basis for assessing a pupil's performance, in relation both to expected attainment and to the next steps needed for the pupil's development.

Both the objectives (attainment targets) and means of achieving them (programmes of study) should leave scope for teachers to use their professional talents and skills to develop their own schemes of work, within a set framework which is known to all. It is the task of the Working Group on Design and Technology to advise on that framework for Design and Technology.

The Design and Technology Working Group forwarded its Final Report in May 1989, and it is encouraging to note that its proposals, together with those already published for Science, and for Primary School Technology, are almost entirely in line with the advice given in *Teaching Craft, Design and Technology: 5–13*, and readers of this book will already be well prepared for the curriculum needs of Design and Technology within those age ranges.

In drawing up recommendations for primary phase Design and Technology the Science Working Party appears to have recognized that scientific knowledge is not entirely transportable to technology, that the application of science is a very different activity from applied science, and that technology involves meeting human needs and draws the knowledge it requires for solving problems from many different disciplines.

For the 5–11 age group the Report lists four attainment targets for Technology:

1. TECHNOLOGY IN CONTEXT
   Children should know that the response to the needs of the living and man-made world has often resulted in a technological solution. They should understand that there can be benefits and drawbacks and realise that this has implications for their own lives, that of the community and the way we make decisions.
2. DESIGNING AND MAKING

Children should be able to design and make an artefact, product or system. They should be able to select and use materials to match specific needs; be able to use tools safely to cut, join and mould them with due regard to aesthetic and functional properties.

3. USING FORCES OF ENERGY

Children should be able to develop and use their knowledge and understanding of forces – both static (in structures) and dynamic (in moving things). They should develop and use their knowledge and understanding of energy, its sources, uses and ways of controlling it.

4. COMMUNICATING TECHNOLOGY

Children should be able to communicate clearly their stages of thinking, designing and making and evaluating using a variety of means such as modelling, drawing, oral or written, mathematical or computer techniques. They should be able to select the most appropriate method for the audience or purpose.

(National Curriculum Report: Science for ages 5–16: DES 1988)

These four attainment targets are intended to make up the sole Profile Component for Technology (5–11) and they should apply to all pupils in the primary school age range.

As this book also covers the first two years of secondary education, with pupils in either middle or secondary schools, mention must be made regarding attainment targets and levels of achievement for this age range although, at the time of writing, the Final Report of the National Curriculum Working Group on Design and Technology has not received final approval from the Seecretary of State for Education and Science.

The Final Report lists four attainment targets for the whole age range covered, and specifies the knowledge, skills and processes which, through the programmes of study, must be taught across a wide range of contexts if the attainment targets are to be achieved. It also recommends a single attainment target for Information Technology (IT) in order to guarantee a minimum entitlement of IT experience for all pupils and a single profile component called 'Design and Technological Capability' reflecting the holistic nature of the pupils' activity in this subject area.

The four attainment targets recommended are:

AT 1. IDENTIFYING NEEDS AND OPPORTUNITIES

Through exploration and investigation of a range of contexts based on home, school, local community, recreation and hobbies and the world of work pupils should be able to identify and report needs and opportunities for design and technological activities.

AT 2. GENERATING A DESIGN PROPOSAL

Pupils should produce a realistic and appropriate design solution within their own capabilities by originating, exploring and developing design and technological ideas and by refining and detailing the chosen ideas.

AT 3. PLANNING AND MAKING

Having developed a plan from their previously developed design solution the pupils should be able to identify, manage and use appropriate resources including knowledge and processes in order to make an artefact, system or environment.

AT 4. APPRAISING

Based on an appraisal of the processes and effectiveness of their own design solution pupils should be able to communicate and act constructively upon the design and technological effectiveness of their own work as well as upon that of others, both contemporary and from other periods and cultures.

Each attainment target is linked to ten levels of achievement reflecting the broad spread of attainment in pupils of the same age and identifying possible strands of progression within the target statement.

The Final Report suggests a separate programme of study for each of the ten levels of attainment but each programme relates to all four attainment targets so that, for example, the programme of study for level one specifies the items, skills and processes which should be taught to pupils to enable them to achieve level one across all four design and technology attainment tartets.

The areas covered by the programme would include:

materials and components	exploring and investigating
energy	imaging and generating
business and economics	modelling and
tools and equipment	communicating
aesthetics	organizing and planning
systems	appraising
structures	health and safety
mechanisms	social and environmental

The programmes of study outlined in both Reports give every opportunity for individual interpretation and for continued freedom in teaching styles and organization but, as with all new developments in education, there is a real danger of regression in that some teachers will try to defend the status quo – often from a narrow subject viewpoint, and there is a similar danger in connection with the moderated assessment/testing procedures proposed in the new legislation. Most primary teachers and, certainly, nearly all infants phase teachers, have never been involved in moderated testing in the past and one of the objectives of the new in-service provision must be to find ways of helping teachers to test and assess objectively and to be able to complete statements of attainment in order to prevent a reversion to stylized examination preparation of the sort which, regrettably, often existed in the days of the 11+ examination. Perhaps the following should be 'writ large' in all school staffrooms:

THE PURPOSE OF ASSESSMENT as an integral and ongoing part of the teaching process is:

* to show and confirm what a pupil has learned and mastered;
* to highlight the need for consolidation and revision;
* to inform decisions about the next steps or stages;
* to enable both teachers and parents to ensure that adequate progress is being made;
* to give other valid and appropriate information.

Earlier in this chapter mention was made of the imminent need to introduce the new curriculum proposals and a study of the proposed implementation timetable will emphasize the urgency of the situation. The latest information available suggests the following time-scale for the introduction of Design and Technology (see Table 10.1).

Because of the close relationship between Science and Technology in the early years of education, it will almost certainly be desirable to develop a reasonably common curriculum pattern for both aspects based largely on local and personal needs and interests and, in most cases, developed from the children's imaginative play, with the teacher acting as guide and enabler. Investigating, understanding, making and doing are all major components of the young child's day and, as has been

Table 10.1 Proposed timetable for implementation of National Curriculum

Date	5 to 7 yrs	7+ to 11 yrs	11+ to 14 yrs	14+ to 16 yrs
Autumn 1990	Attainment targets, etc.		Attainment targets, etc.	
Autumn 1991		Attainment targets, etc.		
Summer 1992	Unreported assessment			
Summer 1993	Reported assessment		Unreported assessment	
Autumn 1993				Attainment targets, etc.
Summer 1994		Possible unreported assessment	Reported assessment	
Summer 1995		Either unreported or reported assessment		GCSE in design and technology*
Summer 1996		Reported assessment (if not in 1995)		

Note
* It is possible that by 1992 there will be a single GCSE in Design and Technology in time for the 1994 examination.

147

said earlier, provided that the teacher encourages and motivates the child, and ensures that there is real progression, then it will not be too difficult for the attainment targets to be achieved at appropriate levels. What will be more difficult at first will be for the teacher to recognize the technological implications of the creative play activity in order to exploit the situation and this may be the time for the reader to have another look at chapter one of this book.

The primary teacher's confidence is not helped by the fact that, to date, there is no real body of knowledge called 'CDT' as there is for most other curriculum areas, and although this implies that any aspect of human need which can be satisfied or improved by any form of technological activity has relevance, it is not easy for the non-specialist teacher to ensure that the pupil is well motivated unless both teacher and pupil can see the relevance of the chosen solution. Greater environmental awareness increases the range of situations and the incentives, thereby helping the pupil to find and identify situations which could lead to interesting and satisfying problem-solving activity, but this also adds to the teacher's problems if there is any doubt as to what is relevant under the heading of Technology. It would seem certain that the drawing up of attainment targets and programmes of study could be a major advance as in future CDT would have an identifiable core content, provided that this is interpreted intelligently and without loss of flexibility. Science is at last breaking free of its traditional stylized approaches and predetermined conclusions – it would be a disaster if its place were to be taken by a fossilized form of CDT.

Most primary and middle years teachers have always been good at encouraging children to 'make and do', although sometimes the work has been too closely led and controlled by the teacher. At least there is a wealth of experience in organizing and controlling a classroom where a variety of group activity, including model-making and other practical work is in progress. This suggests that it will not be too difficult to take on board the programmes of study required for the National Curriculum. However, to date in most primary schools the model-making has been of the static kind, and few pupils have been encouraged to develop, make and evaluate things that work. Under Attainment Target 3 children will be required to investigate, design and make structures and mechanisms which will involve the production, storage, release or control of energy including gravitational, hydraulic, pneumatic, magnetic, electrical or electronic devices, and this goes far beyond the traditional territory of the primary school teacher. However, if the starting point for energy sources is seen in terms of the children themselves, the bicycle pump, the sand hopper, a plastic water container or 'squeezy' bottle, the torch battery and the inclined plane, then the technology involved does not seem quite so frightening. The next step, and a big one for some teachers, will be to move on to control activity including electronic and microprocessor control, and the use of information technology to communicate and retrieve ideas. Now that the majority of primary pupils have access to microcomputers and more and more teachers have acquired appropriate expertise, future activity will include increasing the use of computer graphics and micro-electronic control. By using a 'mouse' for direct screen painting or perhaps by means of a 'joystick', even very young pupils can produce exciting visual display unit (VDU) graphics in connection with problem-solving situations. By using electronic toys the young child can be involved in electronic control while the older pupil can control the stepper motors on his own model via an umbilical cord and interface connected to the microcomputer. Unfortunately, this area of activity is dogged by the belief that electronics is a science simply because electronics is studied as part of a physics course and, to a lesser extent, the same problem exists with hydraulics and pneumatics. In the primary school we must not be too concerned about that area of knowledge which we call physics . . . our concern is all about what the device or artefact can do and how we can use it in our chosen solution. Children use the water tap, the TV set, the video-recorder, and sometimes the microwave oven in their own homes without worrying about how they work, so why should we feel inadequate if we use the same approach in the classroom?

From using very simple approaches in the early years, pupils should progress to a familiarity with the 'Microelectronics-For-All' (MFA) material towards the end of the stages of schooling covered by this book. The principle throughout is to use these newer technologies as enablers adding to the potential of the problem-solving activity, and these, incorporated in models

149

made perhaps using Meccano, Fischer Technik, Lego Technical Functions or Economatics items for any necessary mechanical or electronic components, can lead to exciting and open-ended problem-solving activity and can provide the motivation for scientific enquiry and understanding. The MFA programme mentioned above has an important part to play in the work of the 11 to 13 age group as it uses electronics to give experience in, and to show the relevance of, logical thought. This programme is truly interdisciplinary and brings considerable benefit to later work in all so-called 'subject' areas.

Perhaps a few words of caution could be included here; whilst the use of a 'black box' approach is advocated in order to widen the scope of the pupil's work and enquiry it must be remembered that this is part of the enabling process within problem-solving activity. Some suppliers and manufacturers of electronic and other 'kits' produce material which is simply clipped or plugged together as indicated in the accompanying leaflet and, hey presto! the thing lights up, buzzes or jumps about as promised in the instructions. Considerable thought needs to be given to ways of using this sort of material if the purpose is to be educational rather than just keeping the pupils quiet and amused. As in all CDT activity, the teacher must play a key role in extending the pupil's horizons and standards and in encouraging purposeful investigation and progressive experimentation. It is not necessary to be an expert in electronics to be able to recognize the potential and scope of kits produced by educational manufacturers such as Economatics, Fischer Technik or Unilab, but it is essential for the teacher to ensure that apparatus and equipment made available during problem-solving activity has relevance for that particular line of enquiry and that, should the pupil need help, he or she can be directed to appropriate information or guidance. Unfortunately, as with the primary teacher who started teaching his pupils German on the strength of a five-day visit to a Bavarian beer festival and one term's attendance at a beginner's evening class in the language, so there are teachers who subject their pupils to a watered down 'copy-me' electronics course based on their own very limited experience on a one-day LEA conference. There would appear to be no place whatever for a set course in electronics (or hydraulics or whatever) with pupils in the 5–13 age range, as the aim should be not to teach these activities as formal subjects or packages of tricks but rather that the opportunities presented by this new range of activity should be seen as an addition to the tools, materials, knowledge and processes used in purposeful problem-solving activity.

For some teachers not yet confident in the new technologies, the way ahead in the immediate future may be to exploit the area of alternative technology; that is, by developing activities centred around man's search for technologies which could benefit the Third World and which do not rely on fossil fuels or nuclear energy to provide motive power but which utilize, perhaps, wind, water or muscle power for their energy source. The large wind generators and the hydroelectric power-station built inside a Welsh mountain are examples of possible motivation. However, the two avenues are not mutually exclusive as the enterprising teacher may see possibilities in the use of alternative technology to provide the energy source for, say, an electronic circuit. Many of the fields of interest listed in chapter four have particular relevance for themes of this sort. For example, a study of the machines and devices used in Ancient Greece or by the Romans could lead to problem-solving activity aimed at present-day needs but using simple technological solutions such as might have been possible in those far-off times had there been access to twentieth-century tools and materials. Similarly a study of the machines and devices used in and around the home at the turn of the century (that is, before the availability of electricity in the home) could be a valuable source of situations leading to problem-solving activity and the creation of appropriate working models which lend themselves to true evaluation.

CONCLUSION

As the twentieth century moves towards its close it is to be expected that pupils will become more and more aware of the excitements and mysteries of space travel and exploration together with undreamt of developments in the home environment. At the time of writing these words both the Americans and the Russians are actively preparing to build true space stations to act, amongst other things, as launch pads for further exploration of our solar system and beyond. In our pupils' lifetimes a manned landing on Mars will certainly be effected and other explorations will undoubtedly follow, perhaps to the moons of other planets where the parent body is at present unapproachable. Although to some of us space exploration is not far removed from science fiction, to most of our pupils it is not seen as anything particularly remarkable . . . man has been travelling in space since before they were born. These are the youngsters who have been entrusted to our care so that we can prepare them for a life in a society and environment which will be very far removed from the conditions under which our forefathers lived. Their education deserves all the care and consideration that we can muster. No previous generation of teachers has ever had so much responsibility placed upon it as have headteachers and staff at present in our schools.

Appendix A

TOOL AND EQUIPMENT SCHEDULES FOR CDT

INFANT AND FIRST SCHOOLS

Tools Needed per Mobile Trolley

Item	Number required
bench hook	3
bradawl nos 1 and 2	1 of each
brush and dustpan	1
gent's saw 100mm size	3
glasspaper block (cork rubber)	3
junior hacksaw	2
junior hacksaw blades	25
mallet 100mm head	1
pincers 100mm	2
pin hammer	2
pliers: engineer's 100mm	1
pliers: round nosed 100mm	1
saw: hole	1
screwdriver: cabinet 150mm	2
screwdriver: electrician's 100mm	2
Surform block plane	3
Surform shaper tool	3
tack hammer (model-maker's hammer)	2
wheelbrace	1

HSS drills 3mm, 4mm, 5mm, 6mm	1 of each
bench with woodworker's vice	1
clamp on engineer's vice	1
G-cramp 75mm	

JUNIOR AND 8-12 MIDDLE SCHOOLS

Tools Needed per Mobile Trolley

Abrafile: handled	2
bench hook	4
bradawls nos 1 and 2	1 of each
brush and dustpan	2
centre punch	2
coping saw	4
coping saw blades	50
files: half round 2nd cut: handled 150mm	2
files: half round 2nd cut: handled 200mm	2
files: round smooth: handled 150mm	2
files HSE 2nd cut: handled 150mm	2
files: HSE 2nd cut: handled 200mm	2
files: HSE smooth: handled 150mm	2
files: HSE smooth: handled 200mm	2
G-cramp 75mm	3
glasspaper block (cork rubber)	4
hammer: ball pein (or peen or pane) 250g	1
hammer: pin	2
hammer: tack (model maker's)	1
hammer: Warrington 250g	1
junior hacksaw	4
junior hacksaw blades	50
mallet 100mm head	2
marking gauge	1
marking knife	4
nail punch	1
plain brace 200mm swing	1
improved centre bits 9mm, 12mm, 15mm, 25mm	1 of each
rose bit: round shank	1

rose bit: square shank	1
pincers 125mm	2
pliers: engineer's 125mm	2
pliers: long nosed 100m	1
pliers: round nosed 100mm	1
polystyrene cutter: battery operated	1
saw: dovetail 200mm	2
saw: gent's 100mm	4
saw: hole	1
screwdriver: cabinet 150mm	1
screwdriver: electrician's 100mm	2
screwdriver: electrician's 150mm	1
steel rule 300mm	4
Surform block plane	4
Surform plane file	4
Surform round file	4
Surform shaper tool	4
try square 150mm	4
wheelbrace	1
HSS drills 3mm, 4mm, 6mm	1 of each
countersink bit: round shank	1
four-place middle school bench provided with woodworking and metalworking vices	

9-13 MIDDLE SCHOOLS

Tools Required for Work in Wood

bench brush	6
bench hook	18
bevel: sliding	3
bits: centre 6mm to 25mm	1 set
bits: countersink rose 12mm	1
bits: Jennings 6mm, 7.5mm, 9mm	1 each
bits: Ridgeway metric 6mm to 25mm	1 set
bits: short dowel pattern 9mm	1
brace: ratchet 200mm sweep	2
bradawls: assorted	6
calipers: inside 150mm	1 pair
calipers: outside 150mm	1 pair

carving tools	1 set
chisels: bevel edge 6mm	10
chisels: bevel edge 9mm	10
chisels: bevel edge 13mm	10
chisels: bevel edge 19mm	10
chisels: bevel edge 25mm	10
chisels: mortice 6mm	6
chisels: mortice 7.5mm	2
chisels: mortice 9mm	2
cramp: G 75mm	4
cramp: G 100mm	6
cramp: G 150mm	6
cramp: sash 600mm	6
dowelling jig	1
gauges: cutting	1
gauges: marking	18
gauges: mortice	4
gouge slip: fine carborundum	1
gouge slip: medium carborundum	1
gouge: firmer OC 10mm, 15mm, 20mm	1 each
gouge: scribing 6mm, 12mm	1 each
hammer: claw	1
hammer: pin 100g	2
hammer: Warrington 250g	6
knives: marking	18
mallet: carpenter's 100mm	18
mallet: carver's: beech	2
mitre box	1
oilcan: conical	1
oilstone: fine carborundum	1
oilstone: medium carborundum	1
pincers 150mm	2
plane: smoothing: metal 50mm iron	6
plane: technical jack 350mm	3
punch: nail: sizes A B C	1 each
rubber: cork (glasspaper block)	10
saw: bow 250mm	1
blades for above	10
saw: coping	10
blades for above	100

saw: dovetail 200mm	4
saw: gent's 150mm	3
saw: hand 600mm	1
saw: hole 30mm to 60mm	1
saw: panel 550mm	1
saw: tenon 250mm	18
screwdriver: cabinet 75mm	2
screwdriver: cabinet 100mm	2
screwdriver: electrician's 75mm 100mm	1 each
screwdriver: London 150mm	2
spokeshave: steel: flat	2
Surform block plane	18
Surform planer file	18
Surform round file	18
Surform shaper tool	18
trestle: sawing	2
try square 150mm	18

Machines for Woodwork

bandsaw or jigsaw with no-voltage overload release (NVOR) and keylock switch	1
Sharpedge or whetstone grinder	1
lathe: short bed wood with NVOR complete	1
handturning tools long and strong	1 set 6

Benches for Woodwork

benches: four-place, square	4 or 5
plain cover for benches for design work	4 or 5
woodworking vices: quick release 175mm	16 or 20

Tools Required for Work in Metal or Plastics

Abrafile 200mm	6
Abrafile spring clips	1 pair
acid bath: stoneware 250mm dia. × 125mm	1

aprons: leather small	4
brush: steel wire	1
Bunsen burner	1
calipers: inside 150mm	1 pair
calipers: outside 150mm	1 pair
compass: wing 150mm	1
drill: hand (wheelbrace)	2
drill set with stand 1mm to 13mm × 0.5mm	1 set
files: halfround Bastard: handled 150mm	5
files: halfround Bastard: handled 250mm	5
files: halfround smooth: handled 150mm	5
files: halfround smooth: handled 250mm	5
files: round Bastard: handled 150mm	5
files: round smooth: handled 150mm	5
files: round smooth: handled 250mm	5
files: HSE Bastard: handled 150mm	5
files: HSE smooth: handled 150mm	5
files: HSE smooth: handled 250mm	5
files: square 2nd cut: handled 150mm	5
files: square smooth: handled 150mm	5
files: three square smooth: handled 150mm	5
file carding 50mm wide	1 metre
folding bars 350mm	1 pair
gauge: Vernier caliper with case	1
gloves: Neoprene small	5 pairs
goggles: BS 2092 (2)	6 pairs
hacksaw: adjustable	10
blades for above	20
hacksaw: junior	10
blades for above	20
hammer: ball pein (or peen or pane) 250g	10
hammer: engineer's 750g	1
hammer: planishing 100g	2
head: oval 1.5kg approx	1
head: round 1.5kg approx	1
head: square 1.5kg approx	1
mallets: bossing boxwood 50mm dia.	5
mallets: rawhide 50mm dia.	5
micrometer 0–25mm	1
nippers: cutting electrician's 250mm	1 pair

oilcan: forcefeed	1
pliers: combination 125mm	2 pairs
pliers: flat nose 125mm	2 pairs
pliers: round nose 125mm	2 pairs
punch: centre size B	18
rivet set and snap	2 pairs
riveting kit: pop rivet type	1 kit
rule: rustless 300mm	18
sandbag 250mm dia.	1
saw: piercing	2
blades size 3 and 5 for above	25 each
screwdriver: electrician's insulated	
75mm 100mm 150mm	1 each
scriber	18
snips: curved non-nip 200mm	1
snips: straight tinman's non-nip 200mm	1
soldering iron: 110 volt 65w	2
soldering iron: hatchet 250g	1
soldering iron: straight 250g	1
soldering iron heater with cover	1
square: engineer's 100mm	18
stake: creasing iron 4kg approx	1
stake: half moon 3kg approx	1
stake: hatchet 4kg approx	1
stake: round bottom 4kg approx	1
stake holder to suit above	3
stamps: Imperial alphabet: metal 5mm	1 set
stamps: Imperial figure: metal 5mm	1 set
stocks/dies/taps: BA 0,1,2,3,4,5,6	1 each
vice: hand 100mm	2
vice: toolmaker's 100mm	1
vice grips: fibre 75mm	1 box
vice: 75mm engineer's on stand for	
attachment to middle school bench	8 or 10
wrench grip: Mole small size	1

Forge Work Tools

anvil 50kg	1

anvil stand or wood block   1
flatter: rodded   1
fuller: top rodded   1
hardie   1
hot set: rodded   1
sledge hammer: handled 2kg approx   1
swages: top rodded 12mm   1
tongs: closed mouth flat jaw   1
tongs: hollow bit 6mm   1
tongs: hollow bit 12mm   1
tongs: open mouth flat jaw   1
tongs: square hollow bit 6mm   1
tongs: square hollow bit 12mm   1

Machines for Metalwork and Plastics

bench drill with chuck guard, no-voltage overload
   (NVOL) and foot switch   1
machine vice for above   1
combined brazing hearth and forge   1
grinding machine: double ended with foot switch   1
guillotine notcher: Gabro 2M2 with stand   1
lapidary tumbler with 2 barrels   1
lapidary combination unit   1
lathe: 250mm swing training with NVOL   1
low voltage lighting unit for above   1
model maker's lathe: Unimat with NVOL   1
polishing machine: double ended with foot
   switch   1

Graphics Furniture

Thornton Desra or Oakland drawing frames   4 sets

Appendix B

USEFUL ADDRESSES

British Association for the Advancement of Science,
23 Savile Row, London W1X 1AB

British Standards Institution,
2 Park Street, London W1A 2BS

Cement and Concrete Association,
Education Division, Fulmer Grange, Fulmer,
Slough SL2 4QS

Central Film Library,
Chalfont Grove, Chalfont St Peters, Bucks SL9 9AF

Confederation of Design and Technology Associations
(CODATA),
Hon. Secretary, Brunel University, Shoreditch Campus,
Cooper's Hill, Englefield Green, Egham, Surrey TW20 0JZ

Crafts Council,
12 Waterloo Place, London SW1Y 4AU

Department of Education and Science,
Elizabeth House, York Road, London SE1 7PH

Design Council,
28 Haymarket, London SW17 4SU

Education Service for the Plastics Institute (ESPI)
  Department of Creative Design, Loughborough University,
  Loughborough, Leicestershire LE11 3TU

National Centre for School Technology (NCST),
  Trent Polytechnic, Burton Street, Nottingham NG1 4BU

Appendix C

SUPPLIERS OF MATERIALS AND EQUIPMENT

Benches and Tables

Emmerich (Berlon) Ltd, Wotton Road, Ashford, Kent TN23 2JY

Lervad (UK) Ltd, 4 Denham Parade, Oxford Road, Denham, Bucks UB9 4DZ

E. J. Arnold and Son Ltd, Butterley Street, Leeds LS10 1AX

Drawing Furniture and Equipment

British Thornton Ltd, PO Box 3, Wythenshawe, Manchester M22 4SS

Oakland Design Products Ltd, Telford Industrial Centre, Stafford Park 4, Telford

Quadrant Educational Drawing Equipment Ltd, 61a West Road, London E15 3PX

Drawing Pens and Equipment

Berol Ltd, Berol House, Oldmedow Road, Kings Lynn, Norfolk PE30 4JR

Staedtler (UK) Ltd, Pontyclun, Mid Glamorgan CF7 8YJ

A. W. Faber-Castell Ltd, Crompton Road, Stevenage, Herts SG1 2EF

C. W. Edding (UK) Ltd, North Orbital Trading Estate,
Napsbury Lane, St Albans, Herts AL1 1XQ

Helix International Ltd, PO Box 15, Lye, Stourbridge,
West Midlands

Inscribe Ltd, Caker Stream Road, Mill Lane Industrial Estate,
Alton, Hants GU34 2QA

Rexel Ltd, Gatehouse Road, Aylesbury, Bucks HP19 3DT

Rotring (UK) Ltd, Building One, GEC Estate,
East Lane, Wembley, Middlesex HA9 7PX

Enamelling Equipment and Supplies

W. G. Ball Ltd, Anchor Road, Longton,
Stoke on Trent, Staffordshire ST3 1JW

Craft O'Hans, 21 Macklin Street, London WC2B 5NH

Griffin and George Ltd, Ealing Road, Alperton,
Wembley, Middlesex HA0 1HJ

Catterson Smith Ltd, Tollesbury, Maldon, Essex

General Craft Equipment and Materials

E. J. Arnold and Son Ltd, Butterley Street,
Leeds LS10 1AX

Nottingham Educational Supplies, 17 Ludlow Hill Road,
Melton Road, West Bridgford, Nottingham NG2 6HD

Dryad, PO Box 38, Northgates, Leicester LE1 9BU

Hand Tools

Buck and Hickman Ltd, Gravelly Industrial Park,
Tyburn Road, Birmingham

John Hall Tools Ltd, Educational Division,
3/4 Alston Road, off Portway Industrial Estate, Oldbury,
Warley, West Midlands B68 2RH

Heward and Dean Ltd, 90/94 West Green Road,
Tottenham, London N15 4SR

Neill Tools Ltd, Napier Street, Sheffield S11 8HB

Rabone Chesterman Ltd, Whitmore Street, Hockley, Birmingham B18 5BD

Bahco Products (Record Ridgway Tools Ltd), Educational Service, Parkway Works, Sheffield S9 3BL

Stanley Tools Ltd, Woodside, Sheffield S39PD

PTS Tool Speicalists Ltd, PO Box 242, Henley Street, Camp Hill, Birmingham

Heat Treatment and Forgework

Flamefast Ltd, Pendlebury Industrial Estate, Manchester M27 1FJ

Wm Allday & Co. Ltd, Alcosa Works, Stourport-on-Severn, Worcs DY13 9AP

Vaughans (Hope Works) Ltd, PO Box 2, Hope Street, Dudley, West Midlands DY2 8RD

Plastic Forming Machines

C. R. Clarke & Co., Carregammon Lane, Ammanford, Dyfed SA18 3EL

Griffin & George Ltd, Ealing Road, Alperton, Wembley, Middlesex HA0 1HJ

Electronics and Technology

Economatics Educational Division Ltd, 4 Orgreave Crescent, Dore House Industrial Estate, Handsworth, Sheffield S13 9NQ

Heron Educational Ltd, Unit 12, Kenilworth Works, Denby Street, Sheffield S2 4QL

Surplus Buying Agency for Schools (SBAS), Woodbourn Road, School, Woodburn Road, Sheffield S9 3LQ

National Centre for School Technology (NCST), Trent Polytechnic, Burton Street, Nottingham NG1 4BU

RS Components Ltd, PO Box 427,
13–17 Epworth Street, London EC2P 2HA

Unilab, Clarendon Road, Blackburn, Lancs BB1 9TA

Machine Tools

Denford Machine Tools Ltd, Birds Royd,
Brighouse, Yorks HD6 1NB

Wadkin Ltd, Green Lane Works, Leicester LE5 4PF

Startrite Machine Tool Co. Ltd, 625 Princess Road,
Dartford, Kent DA2 6EH

Boxford Machine Tools PLC, Wheatley, Halifax,
West Yorkshire HX3 5AF

Gabro Engineering Ltd, Hathersham Close,
Smallfield, Surrey RH6 9JE

Elliot Machine Equipment Ltd, BEC House,
Victoria Road, London NW10 6NY

Cowell Engineering Ltd, 95–101 Oak Street,
Norwich NR3 3BP

Appendix D

USEFUL FURTHER READING

General titles

AACDT (1985) *Designing and Making*, CDT Centre, Burley in Wharfedale: Association of Advisers in Design and Technology.

Assessment of Performance Unit (1987) *Design and Technological Activity: A Framework for Assessment*, London: HMSO

Cave, J. (1986) *Technology in Schools*, London: Routledge & Kegan Paul

Coventry Education Authority (1983) *An Introduction to Craft, Design and Technology in the Primary Curriculum*, Coventry: Coventry Education Authority

Design Council (1987) *Design and Primary Education*, London: Design Council

Department of Education and Science (1980) *CDT in Schools: Some Successful Examples*, London: HMSO

Department of Education and Science (1982) *Technology in Schools*, London: HMSO

Department of Education and Science/Welsh Office (1988) *Science for Ages 5–16*, London/Cardiff: HMSO

Kimbell, R. (1982) *Design Education in the Foundation Years*, London: Routledge & Kegan Paul

Luddington, D. (1981) *Design in Technical Studies*, Glasgow: Blackie

Schools Council (1981) *The Practical Curriculum*, London: Methuen Educational

Schools Council (1983) *Primary Practice*, London: Methuen Educational

Shaw, D. and Reeve, J. (1978) *Design Education for the Middle Years*, London: Hodder & Stoughton
Williams, P. and Jinks, D. (1985) *Design and Technology 5-12*, London: Falmer Press
Zanker, F. (1979) *Design and Craft in Education*, Leicester: Dryad

Resource titles

Gibbs-Smith, C. (1978) *The Inventions of Leonardo Da Vinci*, London: Phaidon (Book Club Associates)
Graf, R. and Whalen, G. (1975) *How it Works Illustrated*, London: Souvenir Press
Paladin (1975) *How Things Work (Volumes 1 and 2)*, St Albans: Paladin
Schools Council (1972) *Project Technology Handbook (2) Simple Bridge Structures*, London: Heinemann Educational
Schools Council (1973) *Project Technology Handbook (7) The Ship and Her Environment*, London: Heinemann Educational
Schools Council (1973) *Project Technology Handbook (10) Industrial Archaeology for Schools*, London: Heinemann Educational
Schools Council (1975) *Project Technology Handbook (11) Industrial Archaeology for Watermills and Waterpower*, London: Heinemann Educational
Science 5-13 (1972/4) *Science From Toys (Books 1, 2, 3 and Background)*, London: MacDonald Educational
Science 5-13 (1972/3) *Structures and Forces (Books 1, 2 and 3)*, London: MacDonald Educational
Taylor, G. H. (ed.) (1982) *The Inventions that Changed the World*, London: Readers Digest Association

Tools, processes and materials

Beasley, D. (1984) *Design Presentation*, London: Heinemann Educational
Beasley, D. (1979) *Design Illustration: Sketching and Shading Techniques.* London: Heinemann Educational

Breckon, A. and Prest, D. (1988) *Introducing Craft, Design and Technology*, London: Hutchinson

Gilbert, C. (1988) *Look! Primary Technology*, Edinburgh: Oliver and Boyd

Hicks, G. A., Heddle, G. M. and Bridge, P. A. (1975) *Design and Technology in Metal*, Exeter: Wheaton

Millett, R. (1977) *Design and Technology Plastics*, Exeter: Wheaton

Millett, R. and Storey, E. W. (1978) *Design and Technology Wood*, Exeter: Wheaton

Penguin (1976) *Kites*, Harmondsworth: Penguin

Schools Council (1974) *You Are a Designer*, London: Edward Arnold

Schools Council (1974) *Project Technology Handbook (8) Design with Plastics*, London: Heinemann Educational

Science 5–13 (1972) *Working with Wood (Books 1, 2, and Background)*, London: MacDonald

Science 5–13 (1973) *Metals (Books, 1, 2 and Background)*, London: MacDonald

Science 5–13 (1974) *Children and Plastics (Book 2 and Background)*, London: MacDonald

Science 5–13 (1974) *Science – Models and Toys (Book 3)*, London: MacDonald

Shooter, K. and Saxton, J. (1987) *Design Technology – Making Things Work*, Cambridge: Cambridge University Press

Stokes, P. (1976) *Looking at Materials*, Sunbury on Thames: Nelson

Stokes, P. (1976) *Working With Materials*, Sunbury on Thames: Nelson

Willacy, D. M. (1987) *Craft and Design in Wood*, London: Hutchinson

Willacy, D. M. (1986) *Craft and Design in Metal*, London: Hutchinson

Williams, P. H. M. (1970) *Lively Craft Cards (1) Using Waste Materials*, London: Mills & Boon

Williams, P. H. M. (1971) *Lively Craft Cards (2) Making Musical Instruments*, London: Mills & Boon

Williams, P. H. M. (1971) *Lively Craft Cards (3) Models From Cardboard Tubes*, London: Mills & Boon

Williams, P. H. M. (1972) *Lively Craft Cards (4) Seasonal*

Activities, London: Mills & Boon

Williams, P. H. M. (1971) *Lively Craft Cards (5) Look! It Works!* London: Mills & Boon

Williams, P. H. M. (1973) *Lively Technology Cards (1) Introducting Wood*, London: Mills & Boon

Williams, P. H. M. (1974) *Lively Technology Cards (2) Introducing Metalwork*, London: Mills & Boon

Williams, P. H. M. (1975) *Lively Technology Cards (3) Introducing Glass Reinforced Plastics*, London: Mills & Boon

Yarwood, A. and. Dunn, S. (1986) *Design and Craft*, London: Hodder & Stoughton

Yarwood, A. and Orme, H. (1983) *Design and Technology*, London: Hodder & Stoughton

Health and safety

British Standards Institute (Latest Revision) *BS 4163 Health and Safety in Workshops of Schools and Colleges*, London: British Standards Institute

Department of Education and Science (1981) *Safety in Practical Studies*, London: HMSO

Journals

5–13 Education (monthly), Benn Business Magazines, Maidstone, Kent, ME17 4BR

School Technology (quarterly), National Centre for School Technology, Trent Polytechnic, Nottingham

The Stanley Link in CDT (quarterly), Stanley Tools Ltd, Sheffield

Films and video

Engineering is . . . (suitable for 12+ groups), Rolls Royce, Public Affairs Department, 65 Buckingham Gate, London SW1E 6AT

Practical Thinking, Department of Education and Science,
  Central Film Library
Technology Starts Here, Department of Education and Science,
  Central Film Library

INDEX

Adams, Kenneth 124
adhesives 18–19
aims and objectives of CDT 11–13, 14; for 5–13 age group 13–14
air movement as field of interest 81
alternative technology 151
anemometer as design problem 42
appearance, good in good design 9
appraising as attainment 145
art and craft and technological extensions of subject matter 64, 71
art galleries and resources 122
assessment 124–35; aims of 125; and pupils' profile sheets 128, 132–4; purpose of 146; of skills 127, *see also* profile assessment
Assessment of Performance Unit (APU) 126
Association of Advisers in CDT 11
attainment targets in National Curriculum 143–4
attitudes in design and technology 4
awareness in design and technology 4

balancing in problem-solving situations 53
bench hook 101–2
benches: craft 99; for painting 113; for woodwork in middle schools 157; in workshops 108, 111
bricks in problem-solving situations 55
bridge-climbing vehicle, competition problem 48–9
bridges: in problem-solving situations 57; study of in primary schools 23
British Association for the

Advancement of Science 44

cable climbing, competition problem 50–1
catapaults in problem-solving situations 55
charts 90
classroom: health and safety in 137–8; in middle schools 103; organizing 97–104; as resource 122–3; space for displays in 100; storage system in 98–9, 100; sundries and materials in 103–5
coding system for resources 118, 121–2
communicating technology as attainment target 144
communication skills 85–96; in design brief 95; in evaluation stage 95; and problem-solving processes 88–9; and technology across the curriculum 61
competitions: annual 44–5; examples of 45–52
computers: and communication skills 96; graphics and control 149; and indexing of resources 120
construction: sound in good design 9; and technological extensions of subject matter 63, 64–5, 70
construction kits 105, 149–50
consumable materials, storage of 116
containers for resources 121
convergent stage of problem-solving activity 33
craft bench 99
craft, definition 5–6, 7–8

craftmanship 6
cross-curricular activities/
   studies 140; drawing in 90; in
   middle schools 29
Curriculum 11–16 (HMI, 1977) 4
curriculum, CDT as major area in
   10–11
*Curriculum Statement for the 11–16
   Age Group* (HMI) 127

design: definition 5–6, 8–9; fitness for
   purpose in 8–9; resources for 117
design brief/process:
   communication skills in 95; as
   cyclical process 41; and evaluation
   134; generating as attainment target
   145; in middle schools 27; as
   problem-solving activity 33
Design Council 5, 19
design and technology: attitudes in 4
Design and Technology Working
   Group (DES) 142–3; attainment
   targets (primary school) 143–4
designing and making as attainment
   target 143–4
desks for practical work 100
Dewey, John 31–2
didactic methods 23–4
displays: of pupil's work 115; space
   for in classroom 100; in workshops
   113–14
divergent stage of problem-solving
   activity 33
drawing: as communication skill 86–
   7; as cross-curricular activity 90
Dymo tape 93; and tools 103, 112

economic factors of technological
   extensions of subject matter 64
education: early years, CDT in 15–30;
   traditions in 1
Education Green Paper (1977) 12
Education Reform Bill (1988) 5, 142
electrical tools in workshops 109
electricity supply: and safety 138; in
   workshops 108, 109
electronic kits 149, 150
emergency stop buttons in workshops
   108
energy, using as attainment target 144
English and technological extensions
   of subject matter 64, 71–4
environmental awareness 148

environmental studies: in primary
   schools 20, 23–4; and technological
   extensions of subject matter 64
European Community: and CDT
   140–1
evaluation 124–35; in problem-
   solving activity 37, 42, 43; stage of
   communication skills in 95
examinations, in CDT 126
exploratory methods 23–4

fields of interest 26, 74–84; air
   movement 81; flight 83; industrial
   archaeology 82; in junior schools
   74–5, 77; river crossing 84; wheels
   as 78–80; wind movement as 26–7
fitness for purpose in good design 8–
   9
flammable materials, storage of 116
flight: as field of interest 83; in
   problem-solving situations 53
floating in problem-solving
   situations 53
floors: and safety 138–9; in workshops
   108
flow diagrams 90
forges, tools needed for 159–60
fumes: extraction of in workshops
   107; and safety 139

games in problem-solving situations
   56
GCSE examinations 126
geography: and fields of interest 76;
   and technological extensions of
   subject matter 64, 68–9
girls: in CDT 3; in competitions 52
goggles/safety spectacles 138, 139
graphics: and communication skills
   85–96; and control by computers
   149; furniture for 160
graphs 90

hand tools in workshops 111–12
health education: and technological
   extensions of subject matter 64
health and safety 136–9; and
   classroom organization 138
Health and Safety at Work Act (1974)
   136
*Health and Safety in School
   Workshops* (BSI, 1975) 138
heat treatment processes 107, 108;

tools for 159–60
history and technological extensions of subject matter 64, 69–70
HM Inspectors of Education 136; and CDT in primary and middle schools 15–16
HM Inspectors of Factories 136
holders and containers in problem-solving situations 52–5, 58, 59

indexing of resources 118–20; and computers 120
industrial archaeology as field of interest 82
infant schools 17–20; communication skills in 85–6; hand tools 19; resources needed 18–19; skills in 19; technological vocabulary 93–4; tools for 100–2, 153–4
information: collected in problem-solving activity 36, 40; sources of in workshops 113–14
Information Technology 144
investigation in problem-solving activity 33–5

junior schools: fields of interest in 74–5, 77; tools for 102–3, 154–5

Keith-Lucas, David 2
Koestler, Arthur 86

labels in problem-solving situations 55
language: graphical communication as 85–96; skills in CDT 92–6
lathe tools 112
learning resources 117–24
libraries: and information sources 114; as resource system 121–2
lighting in workshops 107
line alphabet 91
literacy and relative importance of subjects 124

machines: for metalwork 160; for plastics 160; safety of 138; for woodwork 157; in workshops 108
magazine cuttings as resources 121
mapping in primary schools 21
maps and signs in problem-solving situations 53

materials: in classroom 104–5; and technological extensions of subject matter 63
mathematics and technological extensions of subject matter 64, 70
metalwork: machines needed for 160; tools needed for 157–9
'Microelectronics-For-All' 149–50
middle schools 25–30; classroom 103; in competitions 52; cross-curricular activities in 29; practical work in 26; problem-solving situations in 27–8; syllabus constraints in 107; teachers in 2, 148–9; technology across the curriculum 60–1; tools for 102–3, 154–60; workshop organization in 2, 105–16
model boat/water borne vessel, competition problem 46–7
model glider/heavier than air machine, competition problem 45–6
model vehicles in problem-solving situations 54, 56, 57
models 149–50; in primary schools 24–5
modification in problem-solving activity 37
Montessori, Maria 32
motivation: in problem-solving activity 40, 43; in workshops 113–14
museums and resources 122
music and technological extensions of subject matter 70–1
musical instruments in problem-solving situations 55

National Centre for School Technology 2
National Curriculum 5, 140; and assessment 133; and primary school assessment 141–2; timetable for 147
National Curriculum Final Report (1989) 96
National Curriculum Report (1988) 140; assessment and evaluation 124–35; health and safety 136; and learning resources 117; and problem-solving activity 31; and role of CDT 15; 'Science for ages 5 to 16' 96; and technological awareness 60; and technology 1

needs and opportunities, identifying as attainment target 145
newspapers cuttings as resources 121
non-verbal communication 85
numeracy and relative importance of subjects 124

objectives *see* aims
off-cuts, storage of 109, 111
Optical Coincidence Co-ordinate Indexing (OCCI) 120

packaging in problem-solving situations 57
paint: in problem-solving situations 55; in workshops 113
paper in problem-solving situations 53
Piaget, J. 61, 115
picture books in infant schools 18
planning and making as attainment target 145
plastics: machines needed for 160; and safety 139; tools needed for 157–9
Plastics in Schools: Safety and Hazards (Plastics Institute) 139
practical work 26
Primary Education in England (DES, 1978) 16, 20
primary schools 20–25; in competitions 52; as curriculum area 148; environmental studies in 20, 23–4; health and safety in 137; practical work in 26; and subject liaisons 20–1; teachers in 2, 140–1; technological activity in 20; technology across the curriculum 60–1; touch, sense of 24; written work in 24, *see also* junior schools
problem identification in problem-solving activity 35–6, 40
problem-solving activity/processes/situations 52–9; accurate description in 94–5; and communication skills 88–9; convergent stage of 33; and design spiral 8; divergent stage of 33; in middle schools 27–8; process of 31–45
profile assessment 128–33
Profile Component for Technology (5–11) 144

protective clothing 138
pupil's profile sheet 128; for 6-year old pupil 129; for 9-year old pupil 130; for 12-year old pupil 131

resources: art galleries and museums and 122; central area for 120; children's work in 121; classrooms as 122–3; containers for 121; indexing of 118–20; for learning 117–24; recall system 118; storage system for 98–9; for technological awareness 98
river crossing as field of interest 84
rubber-powered wheeled vehicle, competition problem 45

safety: and fumes 139; in workshops 116
Safety in Practical Studies (HMSO, 1981) 137–8
saws and transfer of training 66–7
schools: changes in 3; *see also under* junior; middle; primary
Schools Council project in Design 2
science: and fields of interest 76; and technological extensions of subject matter 64, 71
Science and Technology competitions 44–5
scientific terminology, need for 95
self-evaluation of pupils 127, 134
sequential actions and communication skills 89
sheet material, storage of 109, 111
signalling devices in problem-solving situations 54
silhouettes for tool storage 113
sketching: and communication skills 89; in workshops 113
skills: acquisition of 7; assessment of 127; in CDT 6–7; in infant schools 19; and spatial concepts 125, *see also* communication skills
social aspects of wheels 79
social implications of technological extensions of subject matter 64
solutions: developing in problem-solving activity 37, 40; investigations of in problem-solving activity 36
spatial concepts, skills and attitudes relating to 125

storage in problem-solving situations 54, 55, 58, 59
storage space: for tools 112–13; for unfinished work 115–16; in workshop organization 107, 109
storage system in classroom 98–9, 100
structures in problem-solving situations 57
sundries in classroom 103–4
syllabus constraints in middle schools 107

target, competition problem 47–8
teachers: at evaluation stage 43; courses in CDT 61; and health and safety 136–7; in middle schools 2, 148–9; in primary schools 2, 140–1; role in infant school 17, 19; and self-evaluation of pupils 127; technological awareness in primary schools 20; and technological extensions of subject matter 67–8; training 140–1
team-teaching 107
Technical Studies 2
technological awareness 9, 60–84; and classroom organization 98; and fields of interest 74–84; in primary schools 3–4; resources for in classroom 98
technological competence 9
technological extensions of subject matter 62–74; economic factors of 64; social implications 64
technological vocabulary 93–4, 95–6
technology: across the curriculum 60–84; in context as attainment target 143; definition 5–6, 9, *see also* alternative technology
'tessera' in problem-solving situations 53
timetable for National Curriculum 147
timing devices in problem-solving situations 56
tools: books on 114; correct handling in middle schools 27; for forges 159–60; handling of in primary

schools 24; in infant schools 19, 100–2, 153–4; for junior and middle schools 102–3, 154–5; and materials in good design 9; for metalwork 157–9; in middle schools 154–60; for plastics 157–9; supervision over 99–100; for woodwork 155–7; in workshops 109, 111–13
touch, sense of in primary schools 24
Towards the Middle School (DES Educational Pamphlet No. 57) 25
training: courses 61; teachers 140–1; transfer of 66–7
transport as topic in primary schools 22

'Understanding Design and Technology' (APU) 126

vehicles: study of in primary schools 22–3
vices 99
visual literacy 92

water lifting as design problem 42
weighing devices in problem-solving situations 57
wheeled projector, competition problem 51–2
wheels as field of interest 78–80
wind movement as field of interest 26–7
windmill in problem-solving situations 56
wooden boards, storage of 109, 111
woodwork: benches needed for 157; machines needed for 157; tools needed for 155–7
working environment for CDT 97–116
working surfaces in classroom 100
workshops: and cross-curricular activities 106; extraction of fumes in 107; layout 110; in middle schools 2, 105–16; safety in 137–9; specialist subject areas 106; tools in 109, 111–13, 153–60
written work in primary school 24